PLEASE STAMP DATE DUE, BOTH BELOW AND ON CARD

DATE DUE	DATE DUE	DATE DUE	DATE DUE
PLEASE DO NOT BORROW UNTIL SEP 3 0 2003			

GL-15

Millikan
QC458.S62 K53 2003
Khabibullaev, P. K. (Pulat Kirgizbaevich)
Phase separation in soft matter physics : micellar solutions, microemulsions, critical phenomena

Springer Series in
SOLID-STATE SCIENCES   138

Springer
Berlin
Heidelberg
New York
Hong Kong
London
Milan
Paris
Tokyo

Physics and Astronomy   ONLINE LIBRARY

http://www.springer.de/phys/

Springer Series in
SOLID-STATE SCIENCES

Series Editors:
M. Cardona   P. Fulde   K. von Klitzing   R. Merlin   H.-J. Queisser   H. Störmer

The Springer Series in Solid-State Sciences consists of fundamental scientific books prepared by leading researchers in the field. They strive to communicate, in a systematic and comprehensive way, the basic principles as well as new developments in theoretical and experimental solid-state physics.

126 **Physical Properties of Quasicrystals**
   Editor: Z.M. Stadnik
127 **Positron Annihilation in Semiconductors**
   Defect Studies.   By R. Krause-Rehberg and H.S. Leipner
128 **Magneto-Optics**
   Editors: S. Sugano and N. Kojima
129 **Computational Materials Science**
   From Ab Initio to Monte Carlo Methods
   By K. Ohno, K. Esfarjani, and Y. Kawazoe
130 **Contact, Adhesion and Rupture of Elastic Solids**
   By D. Maugis
131 **Field Theories for Low-Dimensional Condensed Matter Systems**
   Spin Systems and Strongly Correlated Electrons
   By G. Morandi, P. Sodano, A. Tagliacozzo, and V. Tognetti
132 **Vortices in Unconventional Superconductors and Superfluids**
   Editors: R.P. Huebener, N. Schopohl, and G.E. Volovik
133 **The Quantum Hall Effect**
   By D. Yoshioka
134 **Magnetism in the Solid State**
   By P. Mohn
135 **Electrodynamics of Magnetoactive Media**
   By I. Vagner, B.I. Lembrikov, and P. Wyder
136 **Nanoscale Phase Separation and Colossal Magnetoresistance**
   The Physics of Manganites and Related Compounds.   By E. Dagotto
137 **Quantum Transport in Submicron Devices**
   A Theoretical Introduction.   By W. Magnus and W. Schoenmaker
138 **Phase Separation in Soft Matter Physics**
   Micellar Solutions, Microemulsions, Critical Phenomena
   By P.K. Khabibullaev and A.A. Saidov
139 **Optical Response of Nanostructures**
   Microscopic Nonlocal Theory   By K. Cho

Series homepage – http://www.springer.de/phys/books/sss/

Volumes 1–125 are listed at the end of the book.

Pulat K. Khabibullaev   Abdulla A. Saidov

Phase Separation in Soft Matter Physics

Micellar Solutions, Microemulsions,
Critical Phenomena

With 73 Figures and 15 Tables

 Springer

Professor Pulat K. Khabibullaev
Professor Abdulla A. Saidov
Uzbek Academy of Sciences
Heat Physics Department
28 Katartal Street
Tashkent 700135
Republic of Uzbekistan
E-mail: saidov@uzsci.net

Series Editors:

Professor Dr., Dres. h. c. Manuel Cardona
Professor Dr., Dres. h. c. Peter Fulde*
Professor Dr., Dres. h. c. Klaus von Klitzing
Professor Dr., Dres. h. c. Hans-Joachim Queisser
Max-Planck-Institut für Festkörperforschung, Heisenbergstrasse 1, 70569 Stuttgart, Germany
* Max-Planck-Institut für Physik komplexer Systeme, Nöthnitzer Strasse 38
01187 Dresden, Germany

Professor Dr. Roberto Merlin
Department of Physics, 5000 East University, University of Michigan
Ann Arbor, MI 48109-1120, USA

Professor Dr. Horst Störmer
Dept. Phys. and Dept. Appl. Physics, Columbia University, New York, NY 10027 and
Bell Labs., Lucent Technologies, Murray Hill, NJ 07974, USA

ISSN 0171-1873
ISBN 3-540-43890-4 Springer-Verlag Berlin Heidelberg New York

Library of Congress Cataloging-in-Publication Data.
Khabibullaev, P. K. (Pulat Kirgizbaevich) Phase separation in soft matter physics : micellar solutions, microemulsions, critical phenomena / Pulat K. Khabibullaev, Abdulla A. Saidov. p. cm. –(Springer series in solid-state sciences, ISSN 0171-1873 ; 138) Includes bibliographical references. ISBN 3540438904 (acid-free paper) 1. Soft condensed matter. 2. Phase transformations (Statistical physics) 3. Critical phenomena (Physics) I. Saidov, Abdulla A., 1952- II. Title. III. Series. QC458.S62 K53 2003 530.4'74–dc21 2002030665

This work is subject to copyright. All rights are reserved, whether the whole or part of the material is concerned, specifically the rights of translation, reprinting, reuse of illustrations, recitation, broadcasting, reproduction on microfilm or in any other way, and storage in data banks. Duplication of this publication or parts thereof is permitted only under the provisions of the German Copyright Law of September 9, 1965, in its current version, and permission for use must always be obtained from Springer-Verlag. Violations are liable for prosecution under the German Copyright Law.

Springer-Verlag Berlin Heidelberg New York
a member of BertelsmannSpringer Science+Business Media GmbH

http://www.springer.de

© Springer-Verlag Berlin Heidelberg 2003
Printed in Germany

The use of general descriptive names, registered names, trademarks, etc. in this publication does not imply, even in the absence of a specific statement, that such names are exempt from the relevant protective laws and regulations and therefore free for general use.

Typesetting: Data conversion by Springer-Verlag
Cover concept: eStudio Calamar Steinen
Cover production: *design & production* GmbH, Heidelberg

Printed on acid-free paper   SPIN: 10782905   57/3141/di - 5 4 3 2 1 0

Preface

Today the range of objects under investigation in the physics of multicomponent solutions has been considerably extended to include rod-like, rigid-chain and flexible polymer molecules, as well as others with oriented interactions. Due to their complex geometrical form and the increased number of system components in the solutions obtained, various irregular and regular structures may arise, e.g., molecular complexes and associates, micelles, vesicles, liquid-crystal formations, etc., depending on the component concentration and external conditions.

Multicomponent liquid solutions remain the most difficult objects when it comes to describing the processes involved. This is due to the large number of independent variables and the existence of several multiphase separation regions in such systems, allowing several critical points to merge. Moreover, if a solution contains surfactants as components as well as molecules with some specific interactions (e.g., hydrogen-bonded molecules), one may observe a spontaneously emergent, macroscopically homogeneous, optically transparent state with an implicit manifestation of binarity (high dispersity), but nevertheless thermodynamically stable (microemulsions). Thus, in the absence of a macro-phase separation, some 'implicit' phase transitions are observed in the above systems. In these processes, the concentration fluctuation amplitudes, correlation radius, and susceptibility reach a maximum, finite value. The stability boundaries of such systems include not only margins of the real (macro-heterogeneous) phase separation, but also those of the micro-phase separation (micelle formation, microemulsions, liquid crystal and structure phase transitions, specific points, etc.). The appearance of supramolecular structures compels one to take into account their contributions to macroscopic properties of the systems. Undoubtedly, the appearance of such structures is conditioned by the geometry of molecules and interactions between them. However, properties of the described systems are difficult to predict without exact knowledge of the couple interaction potential and, moreover, supercomputers are needed to calculate the states of systems containing 10^4–10^6 molecules.

A unified theory of the liquid state is still under development. The main source of reliable information about this state may be direct experimental studies on the physicochemical properties of liquid mixtures, to determine a model of physically well-founded correlations between their composition and properties. Therefore, the most promising basic research fields in the physics of soft matter are:

- accumulation of experimental data on structure formation processes,
- elucidation of regularities in spontaneous self-assembly,
- a detailed understanding of self-organization processes in vitro and in vivo with the consecutive appearance and separation of hierarchical structures [1] from the original chaos,
- the study of characteristic times of existence and interaction for the structures.

The term 'soft systems' stands for liquid solutions in which the presence of some oriented intermolecular interaction together with features of the geometrical structure of molecules give rise to a degree of orderliness over intermolecular distances.

Multicomponent solutions are considered to be the most promising objects in the study of structure formation processes. The appearance of a certain orderliness in multicomponent liquid solutions leads to the formation of homogeneous regions, even in such weakly mixing components as water and oil. Most frequently, homogenization is achieved by introducing some quantity of surfactants. In this view, microemulsions are dispersions of water and oil formed by adding amphiphilic molecules. The solution is considered to be homogenized due to a sharp decrease in the surface tension, up to 10^{-3} dyn/cm. However, it should be noted that, although the introduction of a surfactant into the water–oil solution is probably the most effective method, it is not the only way of obtaining a homogeneous solution. A transparent macro-homogeneous solution can be obtained by means of co-solvents, for example middle-chain alcohols, when a liquid–liquid phase transition occurs. Here a number of questions arise. How much do the states differ from each other in these two cases? What is the main point in obtaining a homogeneous state: low surface tension, closeness to the phase transition, or a significant change in molecular diffusion? Is it enough to consider one of these factors to describe the properties of multicomponent solutions? What is the structure of a short-range solution like? How do the structural reconstruction kinetics change? Obviously, it is essential to appraise all these factors when studying the thermodynamic stability of microemulsion states.

Thus, phase separations in soft systems exhibit specific features and should be taken into account both in experimental investigations and theoretical descriptions. This is the aim of the present monograph.

Tashkent,
January 2003

P.K. Khabibullaev
A.A. Saidov

Contents

1. Introduction 1
2. Acoustic Spectroscopy of Ideal Solutions 7
   2.1 Density 8
   2.2 Viscosity 8
   2.3 Ultrasound Velocity, Compressibility, and Heat Capacity 12
   2.4 Ultrasonic Absorption 14
3. Phase-Separating Solutions 23
   3.1 Hydrogen Bonds in Solutions
       with Lower Phase-Separation Critical Point 23
   3.2 Phase Diagrams of Phase-Separating Solutions.
       Order Parameter 25
   3.3 Phase Diagrams of Binary Solutions with One Critical Point . 27
   3.4 Binary and Ternary Solutions
       with Closed Phase-Separation Region 27
4. Dynamics of States Close to Critical 37
   4.1 Low Frequency Acoustic Spectroscopy
       of Weakly Absorbing Liquids 37
   4.2 Acoustic Spectroscopy of Critical Solutions
       with Low Sound Absorption 39
   4.3 Acoustic Perturbation and Correlation Radius
       of Fluctuations in the Vicinity of a Critical Point 43
   4.4 Chemical Reactions in Near-Critical States 45
   4.5 Kinetics of Mono- and Bimolecular Reactions
       Close to a Phase-Separation Critical Point 46
5. Physics of Solutions with Double Critical Point 53
   5.1 Theory of Solutions with Double Critical Point 54
   5.2 Rayleigh Scattering of Light 55
   5.3 Dynamics of Near-Critical States of Solutions with a DCP ... 62
   5.4 Shear Viscosity 63
   5.5 Sound Propagation 67

6. Micellization as a Phase Transition ... 71
- 6.1 Conceptual Experiments ... 71
- 6.2 Electronic Structure of Hydrocarbon Chains of Molecules ... 75
- 6.3 Fluctuon Model of Micellization ... 77
- 6.4 Green Function Method ... 80
- 6.5 Huckel's Method of Molecular Orbitals ... 81
- 6.6 Critical Micellization Concentration ... 84
- 6.7 Micellization as a Phase Transition of Finite Type ... 88
- 6.8 Phase Transitions at Micellization in Solutions of Ionic Molecules ... 90
- 6.9 Kinetics of Micellar and Pre-micellar Associations ... 93
- 6.10 Micellization Under Intensification of Molecular Mass Transfer ... 101
- 6.11 Micellization in the Electric Field of Charged Admixtures ... 105

7. Fluctuation Mechanism of Forced Spinodal Decomposition ... 113
- 7.1 Spinodal Decomposition as a Model for Microemulsion Formation ... 113
- 7.2 Non-equilibrium States in Phase-Separating Binary Liquids and External Perturbations ... 119
  - 7.2.1 Variable Electric Field ... 121
  - 7.2.2 Ultrasound ... 122
  - 7.2.3 Thermal Action ... 122
  - 7.2.4 Optothermal Influence ... 123
  - 7.2.5 Noise Field ... 123
  - 7.2.6 Turbulence ... 124
  - 7.2.7 Shear Flow ... 124
  - 7.2.8 Centrifugal Forces ... 125
  - 7.2.9 Stirring ... 126
- 7.3 External Perturbation and Spinodal Decomposition ... 128
  - 7.3.1 Heating of a System Without Stirring ... 129
  - 7.3.2 Heating of a System by Stirring ... 131
- 7.4 Statistical Account of an External Stirring Field ... 133

8. Weak Stirring and Absolute Instability Phenomena ... 139
- 8.1 Singularity in the Heat Capacity in Forced Spinodal Decomposition ... 139
- 8.2 Extending the Region of Absolute Instability ... 142
- 8.3 Initial Stage Kinetics of Forced Spinodal Decomposition ... 145

9. Microheterophase Relaxation State ... 147
- 9.1 Relaxation State Near the Boundary of Absolute Instability Under Weak Perturbation ... 147

9.2 Surface Tension Energy and Heat Absorption Effect 150
9.3 Thermal Relaxation Effects in the Cellular Structure........ 152

10. Transition from Emulsion to Microemulsion 155
10.1 Microemulsion Structure 155
10.2 Polychronal Relaxation Processes
and Dispersion on the Interface 158
10.3 Stable Microheterophase State
on the Interface of Weakly Dissolved Liquids 161
10.4 Conclusion 166

References 169

1. Introduction

The processes that generate the microemulsion state are complex and versatile. This explains why no exact definition has yet been found for the term microemulsion. It is a transparent solution of a colloidal-disperse type containing water, oil and surfactants. Nevertheless, a microemulsion is indeed generated spontaneously and this may be the reason why researchers have shown such minor interest in the mechanisms underlying its formation.

The increase in the number of components in water–amphiphile–oil–alcohol solutions results in a complex phase behavior. The question of thermodynamic stability when fluid transparent isotropic phases arise is not an easy issue, because the system is complex and microemulsions are structured media whose structure depends on both their molecular interaction and their composition. Moreover, the structure is determined by the microemulsion stability because the dispersion entropy and low surface tension provide an additional contribution to the free energy of the system. It should be noted that structure formation takes place both in the microemulsion itself with its division into layers containing pseudo-phases and in the layers themselves with their division into domains of various types:

- water microdomains formed by water and some amount of alcohol,
- organic microdomains formed by oil, amphiphile and water,
- interface microdomains almost wholly composed of amphiphile and co-surfactant.

Evidently, this is the case when accidental self-organizing objects are under initial conditions, while the chaos is not completely disordered but of a rather dynamical nature [2].

Structure formation resulting from dynamical reorganization is part of a complicated problem associated with the genesis of partially ordered states from disorder. A question arises in this connection: how does a well-organized, structured distribution of the system happen to arise with the initially accidental arrangement?

This question can be partly answered via the results of computer simulation experiments based on the Ginzburg–Landau complex equation describing a very long one-dimensional system, and partly via laboratory experimental data. By means of a random number generator, the initial disorder is set and the spatial dimension is determined as a function of time. Its value begins by decreasing rapidly but then tends to a constant. If the initial conditions are close to sinusoidal, the dimension value after transition tends to the same constant. Under completely different initial conditions, space–time chaos sets in here. Its snapshots exhibit quasi-disorder of a quite definite dimension, which may be described by a finite-dimensional dynamical system.

Laboratory experiments with liquids involve extended systems and the over-criticality slightly exceeds the threshold for formation of a periodic convective structure. The system can be made completely chaotic by simple mechanical stirring. Its further evolution depends on the over-criticality degree. In the case of large over-criticality, one can observe complex non-stationary regimes including those of turbulent convection. Small over-criticality may result in a regular multifaceted lattice of domains divided by boundaries. Computer simulation data and laboratory experiments show that structure formation of this type is, to some extent, a refined order arising from the initial chaos. Undoubtedly, this process can be viewed as self-organization of stochastic structures.

Microemulsion structurization makes it possible to use a pseudo-phase model to study its physical properties. Having a limited set of experimental data, one can calculate compositions of pseudo-phases, reveal so-called pseudo-three-phase planes in the phase space of four-component systems, and thereby efficiently reduce the number of components [3].

The obligatory character of the low-surface tension in the microemulsion follows organically from the concept of structurization. Although the mechanisms that give rise to low surface tension have been under study for a long time, no profound understanding has yet been achieved.

The results from [4] allow us to say confidently that the mechanisms giving rise to low surface tension on oil–water or microemulsion–oil interfaces are different. Whereas in the first case this surface tension is usually attributed to the presence of an amphiphile monolayer with a high surface pressure [5], in the second case various mechanisms are under discussion, ranging from the presence of liquid-crystal layers on the interface boundaries [6] through to the closeness of the system state to critical points [7–9].

In chemical equilibrium, the concentration of amphiphile molecules on the phase-separation boundary is larger than in the bulk. The thermodynamic forces that push a surfactant out to the surface produce surface tension. If σ is the surface tension of an amphiphilic solution, $\sigma_0 - \sigma = \pi > 0$ and, since the function $\partial \mu / \partial x$ (where x is the amphiphile concentration) is positive, then $\partial \pi / \partial \mu$ will also be positive. Therefore, the surface tension decreases with the formation of films and the concentration x in the solution increases. Further, as the concentration exceeds the critical level, the chemical potential of an amphiphile molecule within a micelle becomes less than that of a separate molecule. Such an increase of concentration x above the critical value gives rise to the growth of micellae. In this case the surface tension must be constant. However, it often continues to decrease even for a noticeable excess of the critical micellization concentration (CMC) [10]. Consequently, researchers associate low surface tension not only with the presence of an amphiphilic film, but also with the presence of some critical phenomena. The critical behavior of a system is known to influence the value of the interface tension. According to [4], if ϵ is the distance to the critical point, i.e.,

$$\epsilon = \frac{|Z - Z_{\mathrm{cr}}|}{Z_{\mathrm{cr}}},$$

where Z is the field variable and Z_{cr} is its value at the critical point, then

$$A = A_0 \epsilon^\alpha,$$

where A_0 is the scale factor and α the critical index.

In a number of works [4,7,11,12], the normalized salinity concentration of the solution, or the ratio of water and surfactant concentrations, is used as a field variable. Actually, when the salinity has a certain value, a phase conterminous to microemulsion grows turbid before disappearing. A detailed study of such behavior was carried out in [13]. The measured correlation length of concentration fluctuations proved to be comparable with the dimensions of the microemulsive drops. This means that the critical salinity was still far from the critical point, although the state of the system as a whole could be considered as pseudo-critical. Hence, the physical properties of microemulsions close to critical points may help to understand the mechanisms behind their formation and stability.

In accordance with the concept of invariance [14,15], the critical indices characterizing behavior of physical values for different systems relating to one universality class are equal. The universality class is defined by space dimension and transition order parameter. The universality class for micellar solutions and microemulsions can be defined via their critical indices. Experimental data show that the critical points of pure liquids and binary solutions belong to the same universality class as those predicted by the Ising model for magnetic materials.

With regard to microemulsions, the situation is somewhat different. Actually, some experimental data [12,16] are in agreement with those obtained for pure liquids, whilst others evidence a more complicated behavior [17,18] in microemulsions. For many amphiphilic solutions, the exponents measured do not correspond to the Ising exponents and moreover, in a number of cases they depend on amphiphiles. This is rather clearly seen in the monophase region. Evidently, the critical phenomena occurring in these micellar solutions do not relate to the simple type of the Ising model. The reason for this may lie in the dynamical character of microstructures.

In order to develop a descriptive model of critical behavior, one should understand the phase redistribution mechanism. Regions of microemulsions rich in water or oil are considered to contain oil and water drops, respectively. Drops are supposed to be covered by a film of amphiphile–alcohol mixture whose molecules rapidly interchange through the interface film, i.e., an interdrop exchange between molecules and amphiphiles can take place via the surrounding drop solution. The microemulsion components are redistributed among drops rather quickly due to collisions accompanied by their temporal merger and further decay, as well as partial destruction of drops and their further coagulation into new drops. These processes are determined by the bimolecular exchange rate constant [19]. This is a complex quantity since, besides drop collisions, the exchange reaction involves drop merging as well, i.e., 'dimer' formation, followed by diffusion and reaction between reagents. Here, the reaction rate is limited by destruction of drops.

The lifetime of a transitive dimer may significantly influence the phase separation. Since the lifespan of a dimer becomes longer as molecular interaction increases, other drops can merge with it during its lifetime that will destroy polydispersity and finally give rise to the phase separation of a system. The dependence of the bimolecular exchange rate constant on the level of intermolecular interaction is proved by the sensitivity of the latter to the length of alcohol molecules and oil chains [20].

On the other hand, data on microemulsion dynamics [21] show that microemulsions resemble a critical system more and more closely as the exchange constant increases. Hence, the reason for the phase separation and emergence of critical behavior may lie in drops approaching one another as a result of intermolecular interactions.

The problem of molecular interactions has been considered in [4, 23, 24], which dealt with microemulsions having some excess of oil. The interaction between drops was defined by the van der Waals attraction and screened electrostatic repulsion. In the joint microemulsion–water region, the repulsion forces were spherical. Attractive interactions can favor interpenetration of drops and exchange of their nuclear contents. When the oil concentration in bulk reaches a certain value, percolation phenomena may occur in the system. In consequence, the resulting phase would be rather a bicontinuous structure than a dispersion of isolated drops. When applied to such microemulsions, the ultrasound absorption method failed to reveal fast composition fluctuations similar to those in close-to-critical binary systems [25]. However, some specific characteristics of ultrasound absorption were observed there, which accompany the critical phenomena caused by increase in salinity of the system [26].

All this suggests that the critical behavior of microemulsions and mixtures of simple molecules is probably quite different. It should be noted, however, that critical indices in liquids depend on the direction of approach to the critical point. In this case the solutions under study are systems with a lower critical point, and phase separation comes about when the temperature goes up. Hence, a critical point is usually reached through an increase in the temperature, which is in this case a field variable.

For binary solutions, measurements in one-phase regions can be conducted along the path on which their composition does not change (constant density). If the critical point is set along another direction, e.g., at constant temperature, there should be other exponents. Therefore, it is of great importance to reach the critical point experimentally by changing a field variable under constant temperature and pressure. The main difficulty here is to find an appropriate field variable, which could be controlled throughout experiments and influence the condition of phase separation. In the case of microemulsions, this is not an easy task.

The present monograph attempts to consider the following aspects of the microemulsion state:

- stability of multicomponent systems,
- appearance of lower surface tension, anomalies due to the system being close to phase separation and structure formation,
- dimensions of larger-than-molecular structures, including the kinetics of their formation, destruction and re-organization, and the effects of external influences on their kinetics and stability, etc.

Complication of the near-order structure can help us to discover the correlations between its properties. With this aim in mind, the monograph focuses on systems that have gradually complicated near-order structures. These are binary solutions with ideal solubility (Chap. 2), solutions with structural phase transitions (Chaps. 3, 4, 6), micelle-forming solutions (Chap. 6), solutions with critical phase-separation points (Chaps. 3–5), multicomponent solutions with merging phase transitions (Chap. 5), and microemulsions (Chaps. 1, 7–10).

2. Acoustic Spectroscopy of Ideal Solutions

In this chapter we present data from investigations into the properties of normal solutions. Normal solutions are usually thought to consist of rather simple components, whose molecules are not associated or weakly associated. Among normal solutions, ideal mixtures are of special importance. By definition, they must satisfy the following criteria: $\Delta V = 0$; $\Delta H = 0$; $\Delta S = R \sum x_i \ln x_i$, where ΔV, ΔH, ΔS are the deviations from the additive values of volume, enthalpy and entropy respectively, and x_i is the mole fraction of the ith component.

Hence, ideal mixtures are characterized by athermal properties ($\Delta H = 0$), additivity of volume (density) and molar heat capacity. Liquid mixtures with rather similar molecular structures have properties close to ideal. In most cases, the dependence of thermophysical properties on composition may significantly differ from the simple additivity rule.

Complete information on the thermophysical properties of solutions cannot be obtained from experiment alone, as the combinatorics of the concentrations and temperatures is virtually unlimited. Therefore, it is necessary to develop methods for computing the missing data by revealing functional connections and correlations employing modern theories. For this purpose, we shall discuss the main regularities in the behavior of the density ρ, shear viscosity η_s, surface tension coefficient σ, sound velocity v_c and ultrasonic absorption coefficient α, adiabatic compressibility β_S, volume viscosity η_v, and other characteristics of normal solutions which depend on the composition and temperature.

Physicochemical quantities associated with liquid solutions are usually calculated theoretically using semi-empirical and empirical correlations based on a model representation of the microcrystalline liquid structure of the liquid. The essence of these representations is the assumption that a liquid or a liquid solution consists of separate small regions of 'crystallite' order that retain the properties of a liquid and are located chaotically with respect to each other. These regions are unstable, decaying and forming again, as confirmed by the results of X-ray studies. Thus, a binary solution can be represented as a medium that is a mixture of crystallites of AA, BB, and AB type with the internal structure corresponding to that of the original components A and B and their mixture. If the energies of intermolecular interaction of homogeneous and heterogeneous components are equal, their molecules are distributed in a microscopically uniform way. The properties of such an ideal solution can be obtained by additive summation of the properties of AA, BB, and AB.

2.1 Density

Density is one of the most important characteristics of liquid mixtures. Its value depends on intermolecular interaction forces and molecular structure, so knowing the density makes it possible to judge the interaction forces and changes in molecular structure of liquid systems. In the framework of the microcrystalline model, the density ρ of a binary solution is described by the following expression [27]:

$$\rho = \rho_A X_V^2 + \rho_B (1 - X_V)^2 + 2\rho_{AB}(1 - X_V) , \qquad (2.1)$$

where X_V is the volume concentration of the original components.

The value of ρ_{AB} depends on the interaction forces between molecules of solution components and may be found either by special calculations or from experiments. For an ideal solution, when the bulk effect of formation and decay of the complex AB is absent, the expression (2.1) takes the form

$$\rho = \rho_A X_V^2 + \rho_B (1 - X_V)^2 . \qquad (2.2)$$

Density values measured in experiments designed to study the dependence on composition at various temperatures differ from those calculated by the additivity rule. The difference is most noticeable for solutions whose components diverge significantly with respect to molecular weight and structure. For instance, for benzene–hexadecane solutions, the difference is as much as a few percent.

For solutions whose components are similar by their molecular weight and other parameters, experimental density values almost coincide with those obtained by the additivity rule. For instance, for normal heptane–nonane, nonane–tridecane, tridecane–hexadecane solutions, whose components have similar molecular weights, calculation by (2.2) gives quite a satisfying accuracy up to 0.1%. If a solution contains components with quite different molecular weights and, especially, heterogeneous molecules (see Fig. 2.1), (2.1) yields better results.

2.2 Viscosity

Experimental measurements of viscosity have shown that, on applying trivial additivity, calculated values differ noticeably from the experimental. Therefore, this rule cannot be adopted even in preliminary calculations of the viscosity in liquid mixtures. A number of empirical methods for determining

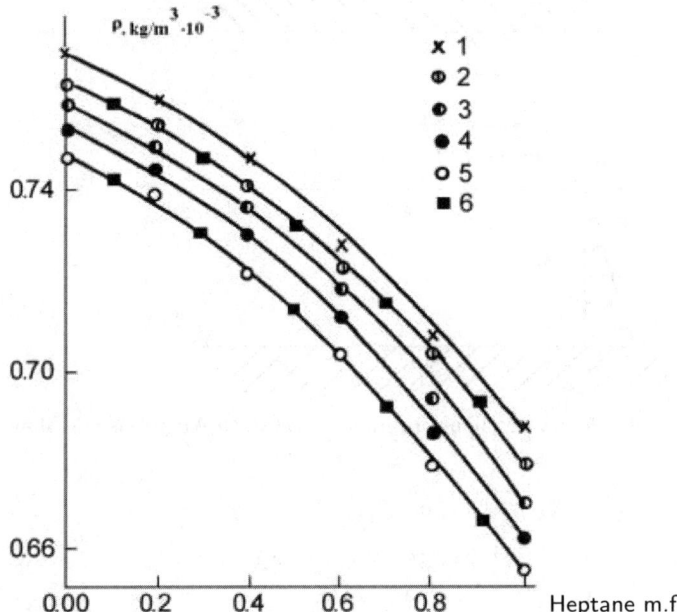

Fig. 2.1. Density curves of a heptane–hexadecane solution for various temperatures: (1) 293 K, (2) 303 K, (3) 313 K, (4) 323 K, (5) 333 K, (6) data at atmospheric pressure

the viscosity of liquid solutions are given in [29–32]. However, as calculations showed, these methods also proved to be too approximative in describing the concentration dependence of the liquid mixtures we have investigated. A calculation method is presented in [33] that makes it possible to determine the viscosity of binary solutions by making use of the viscosity values of pure components as original parameters. The derivation of the formulas was also based on a model assuming the liquid structure to be microcrystalline, i.e., assuming the existence of ordered regions in the liquids. When a system of this type undergoes viscous flow, these regions can behave as solids and then flow only occurs in inter-crystallite regions. In addition, a binary solution is assumed to have crystallites of types AA, BB, and AB. The viscosity of the system is obtained from the following condition:

$$\eta = K_1 \eta_{AA} + K_2 \eta_{BB} + K_3 \eta_{AB} , \qquad (2.3)$$

where $\eta_{AA} = \eta_A$, $\eta_{BB} = \eta_B$, and $\eta_{AB} = \eta_{BA}$ are contributions to the effective viscosity conditioned by viscous friction of homogeneous and heterogeneous regions (crystallites). The coefficients K_i are the probabilities of touching pairs AA, BB, AB. They can be calculated from the condition for uniform distribution of the components in the bulk of the solution:

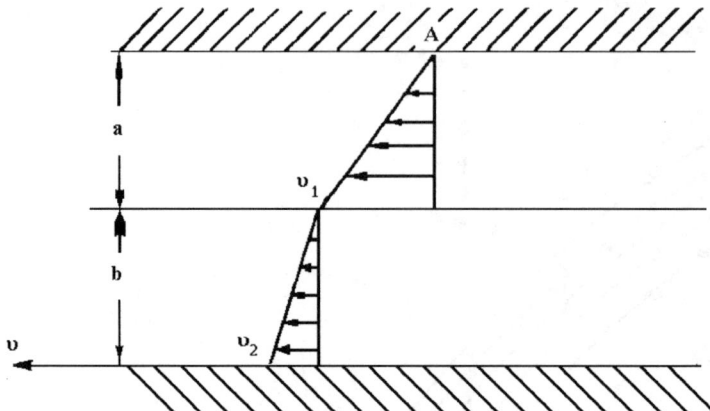

Fig. 2.2. Distribution of velocity gradients in micro-crystallite AB in viscous flow

$$K_1 = X_V^2, \quad K_2 = (1 - X_V)^2, \quad K_3 = 2X_V(1 - X_V). \tag{2.4}$$

The quantity η is determined by simple relations (see Fig. 2.2):

$$F = \eta_A \frac{V_1}{a}, \quad F = \eta_B \frac{V_2}{b}, \quad F = \eta_{AB} \frac{V_1 + V_2}{a + b}, \tag{2.5}$$

where V_1, V_2 are the relative velocities of viscous flow. From (2.5), we obtain

$$\eta_{AB} = \frac{\eta_A \eta_B (a + b)}{a \eta_B + b \eta_A}. \tag{2.6}$$

Supposing $a \sim 1/\eta_A$, $b \sim 1/\eta_B$, we then have

$$\eta_{AB} = \frac{\eta_A \eta_B (\eta_A + \eta_B)}{\eta_A^2 + \eta_B^2}. \tag{2.7}$$

Using (2.4)–(2.7) to calculate the viscosity in a binary solution, we obtain

$$\eta_{\text{eff}} = X_V \eta_A^2 + (1 - X_V)\eta_B^2 + 2X_V(1 - X_V)\frac{\eta_A \eta_B (\eta_A + \eta_B)}{\eta_A^2 + \eta_B^2}. \tag{2.8}$$

Comparison of experimental data with data calculated using (2.8) shows that the latter describes the viscosity concentration dependence in paraffin–paraffin and aromatic–aromatic solutions with accuracy up to 1–3%, whilst for paraffin–aromatic solutions, the experimental and calculated data differ significantly, by more than 10%. We can assume that this is related to the properties of the molecular structure of the solutions, because of the diverse molecular sizes of components and, hence, crystallites. In other words, a component with crystallites of lesser volume V should make a relatively higher contribution to the resulting viscosity because of their large common touching area. This peculiarity may be taken into account by introducing a correction. Let $\eta_i^* = \eta_i K_i$ and

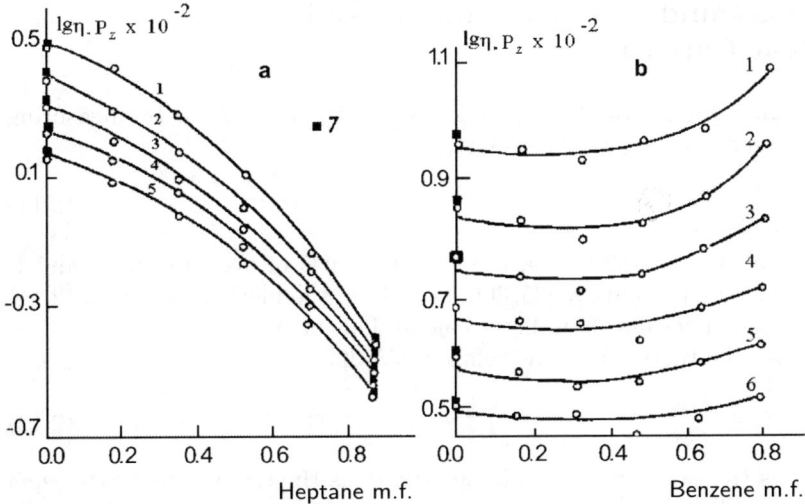

Fig. 2.3. The viscosity curves of (**a**) heptane–hexadecane and (**b**) benzene–octane mixtures for various temperatures: (1) 293 K, (2) 303 K, (3) 313 K, (4) 323 K, (5) 333 K, (6) 343 K, (7) data (30) at atmospheric pressure

$$K_i = \left(\frac{V_A}{V}\right)^{2/3} \approx \frac{\sum \frac{\mu_i}{\rho_i} x_i}{\mu_i/\rho_i}, \qquad (2.9)$$

where η_i^* is the actual viscosity of the ith component in the solution, η_i is its original viscosity, V_A is the average volume, and V is the volume per molecule. Taking into account the expression (2.9), equation (2.8) takes the form

$$\eta_{\text{eff}} = K_1 X_V^2 \eta_A + K_2 \eta_B (1 - X_V)^2 \qquad (2.10)$$
$$+ 2 X_V (1 - X_V) \frac{K_1 K_2 \eta_A \eta_B (K_1 \eta_A + K_2 \eta_B)}{(K_1 \eta_A)^2 + (K_2 \eta_B)^2}.$$

In the form (2.10), the equation for binary solution viscosity was applied to describe the viscosity concentration dependence in liquid mixtures. Calculation errors for the viscosity in paraffin–paraffin systems (Fig. 2.3a) and aromatic–aromatic systems were not more than 1.2% and in aromatic–paraffin systems, 3–5% (Fig. 2.3b).

The proposed method for determining the viscosity in binary solutions of the indicated type has an advantage over those given in the literature as it needs no experimental data for a whole range of concentrations and temperatures, except for the viscosity and density of pure components.

2.3 Ultrasound Velocity, Compressibility, and Heat Capacity

The adiabatic compressibility β_S is known to be determined by measuring the ultrasound velocity ϑ_c. The relation

$$\beta_S = \frac{1}{\rho \vartheta_c^2} \tag{2.11}$$

is a fundamental theoretical basis for making use of acoustic data, and is observed with high accuracy [35, 36]. Therefore, the method based on (2.11) gives more exact results than direct calculations of β_S.

If we use the thermodynamic relation [38, 39]

$$\beta_T = \beta_S + \frac{T\alpha_t^2}{C_P}, \tag{2.12}$$

where β_T is the isothermal compressibility, α_t is the temperature expansion coefficient, and C_P is the heat capacity at constant pressure, or if we take into account that

$$\beta_T = \frac{1}{\rho}\left(\frac{\partial \rho}{\partial P}\right)_T, \quad \left(\frac{\partial \rho}{\partial P}\right)_T = \frac{1}{\vartheta_c^2} + \frac{T\alpha_t^2}{C_P \rho}, \tag{2.13}$$

then there is an obvious possibility for using the data on ultrasound velocity to determine a large number of thermodynamic parameters. It should be noted that when the ultrasound propagates, the medium is stationary but out of equilibrium. Hence the ultrasound velocity ϑ_{c0} of minimal (zero) frequency is a thermodynamically determined value. The ultrasound velocity ϑ_c at any other frequency depends on relaxation times and other characteristics of the process considered. At a frequency of 106 Hz, however, ultrasound propagation in liquids is weakly non-equilibrium, so at these frequencies it may be taken as equal to $\vartheta_c = \vartheta_{c0}$. Experiments show that $\Delta\vartheta_c = \vartheta_c - \vartheta_{c0}$ does not exceed one tenth of ϑ_c.

To calculate C_P by the above method, in addition to the ultrasound velocity data, we also used adiabatic compressibility data [39, 40] and the experimentally obtained value of the thermal expansion coefficient. The Poisson coefficient γ and heat capacity at constant volume C_V were determined by the expression

$$\gamma = \frac{\beta_T}{\beta_S} = \frac{C_P}{C_V}. \tag{2.14}$$

The results for calculation of C_P, C_V and γ are given in Table 2.1.

Using (2.14), we calculated the adiabatic compressibility β_S for all investigated temperatures and concentrations. The isotherms β_S for three types of solutions are shown in Fig. 2.4. They are seen to be nonlinear. This nonlinearity is amplified in systems with heterogeneous components. In benzene–n-hexadecane solutions, β_S goes through a clear maximum. In all solutions, β_S decreases as the temperature increases.

Table 2.1. Values of the heat capacities C_P and C_V [J/mol] and the Poisson coefficient γ

T [K]	C_P (exp.)	C_V	γ
n-tridecane			
303	340.9 ± 10.3	273.6 ± 13	1.246
313	358.3 ± 10.8	291.3 ± 15	1.230
323	364.7 ± 10.9 (385.6 [36])	297.7 ± 15	1.225
333	347.1 ± 10.5	280.4 ± 14	1.238
343	330.8 ± 9.9	265.0 ± 13.2	1.248 ± 0.02
n-hexadecane			
303	432.2 ± 13	349.1 ± 17.4	1.238
313	423 ± 12.6	339.5 ± 17.0	1.246
323	459.6 ± 13.8	375.8 ± 18.7	1.223
333	490.8 ± 14.7	406.43 ± 20.3	1.208
343	468.4 ± 14.1	385.8 ± 19.2	1.214 ± 0.02
Toluol			
293	122.7	87.89	1.396
303	131.9 ± 4.0	98.84 ± 4.8	1.362
313	145.4 ± 4.3	109.9 ± 5.5	1.323
323	152.6 ± 4.6 (162.6 [36])	117.6 ± 5.8	1.298
343	146.6 ± 4.3	112.3 ± 5.6	1.305 ± 0.02

In [35], the adiabatic compressibility of an ideal liquid mixture is shown to be connected with that of the mixture components by

$$\beta_S = \beta_{S(A)} X_V + \beta_{S(B)}(1 - X_V), \tag{2.15}$$

if the corresponding values of C_P/C_V are close to each other. Introducing the so-called interaction constant, which is in this case the compressibility of interacting components $\beta_{S(AB)}$, gives

$$\beta_S = \beta_{S(A)} x_V^2 + \beta_{S(B)}(1 - x_V)^2 + 2\beta_{S(AB)} x_V(1 - x_V). \tag{2.16}$$

Data are given in Table 2.2 for several solutions, together with the values used for $\beta_{S(AB)}$.

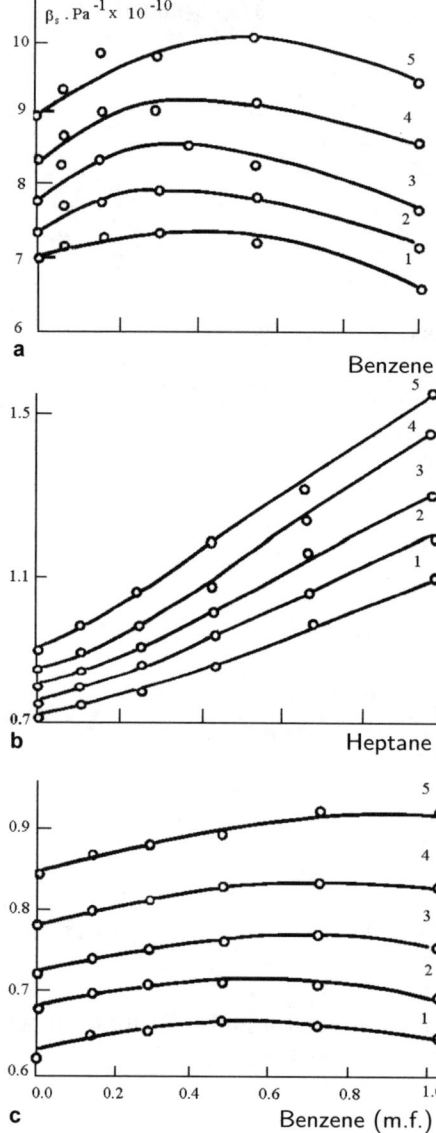

Fig. 2.4. Adiabatic compressibility curves for (**a**) benzene–hexadecane, (**b**) n-heptane-n-hexadecane, and (**c**) benzene–tert-butylbenzene mixtures at various temperatures: (1) 293 K, (2) 303 K, (3) 313 K, (4) 323 K, (5) 333 K at atmospheric pressure

2.4 Ultrasonic Absorption

Investigations of ultrasound wave absorption in liquids make it possible to study their non-equilibrium thermodynamic properties, e.g., bulk viscosity, relaxation times, relaxation fraction of heat capacity.

The results of experimental studies on the coefficient of ultrasonic absorption divided by the squared frequency α_p/ω^2 are given in Fig. 2.5. The

2.4 Ultrasonic Absorption

Table 2.2. Experimental (I) and theoretical (II) values of adiabatic compressibility β_S [10^{-10} Pa^{-1}]

X [m.f.]	293 K		303 K		313 K		323 K		333 K	
	I	II	I	II	I	II	I	II	I	II
Benzene–hexadecane										
$\beta_{S(AB)}$	7.963		8.835		9.859		10.215		11.255	
0.2	7.141	7.084	7.702	7.143	8.302	8.048	8.652	8.604	9.313	9.248
0.4	7.256	7.223	7.678	7.758	8.305	8.357	9.009	8.883	9.782	9.605
0.6	7.304	7.342	7.849	7.963	8.461	8.663	8.966	9.166	9.711	9.979
0.8	7.088	7.315	7.684	8.004	8.148	8.777	9.120	9.300	9.964	10.204
Benzene–tert-butylbenzene										
$\beta_{S(AB)}$	6.838		7.311		7.960		8.406		9.357	
0.2	6.500	6.487	6.966	6.976	7.512	7.436	7.952	8.039	8.781	8.692
0.4	6.595	6.575	7.079	7.067	7.543	7.584	8.184	8.157	8.796	8.902
0.6	6.611	6.634	7.108	7.137	7.652	7.704	8.334	8.273	8.979	9.102
0.8	6.605	6.630	7.136	7.160	7.705	7.761	8.464	8.375	9.313	9.266
Heptane–n-hexadecane										
$\beta_{S(AB)}$	8.398		8.912		9.470		10.076		10.622	
0.2	7.323	7.286	7.728	7.716	8.143	8.176	8.765	8.772	9.419	9.359
0.4	7.751	7.737	8.221	8.222	8.779	8.740	9.423	9.373	10.008	9.991
0.6	8.342	8.371	8.982	8.948	9.623	9.569	10.313	10.282	10.941	10.982
0.8	9.249	9.319	10.002	10.057	10.832	10.852	11.641	11.710	12.491	12.604

complex dependence of α_p/ω^2 on temperature in pure paraffins (Fig. 2.6) has been revealed experimentally. The temperature coefficient $\partial(\alpha_p/\omega^2)/\partial T$ is positive for lower homologues and negative for higher ones. Analogous regularities are observed in a homologous series of aromatics (Fig. 2.7).

16   2. Acoustic Spectroscopy of Ideal Solutions

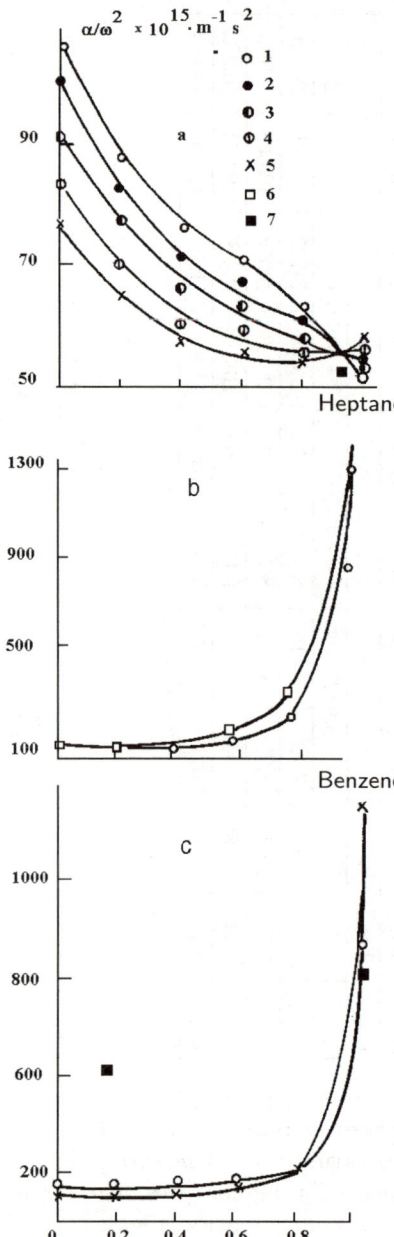

Fig. 2.5. Ultrasonic absorption coefficient curves of (**a**) heptane–hexadecane, (**b**) benzene–octane, and (**c**) benzene–tert-butylbenzene mixtures at various temperatures: (1) 293 K, (2) 303 K, (3) 313 K, (4) 323 K, (5) 333 K, (6) 343 K, (7) data from [42] at atmospheric pressure

At first $\alpha_\mathrm{p}/\omega^2$ is seen to decrease when m increases, reaching a minimum, and then increasing again. This may be supposed to take place due to the two competing types of relaxation processes that occur in such systems.

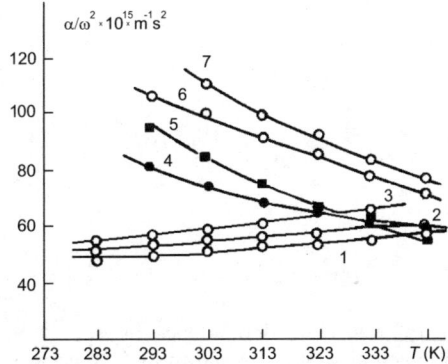

Fig. 2.6. Ultrasonic absorption coefficient curves of (**a**) heptane–hexadecane, (**b**) benzene–octane, and (**c**) benzene–tert-butylbenzene mixtures at various temperatures: (1) 293 K, (2) 303 K, (3) 313 K, (4) 323 K, (5) 333 K, (6) 343 K, (7) data from [42] at atmospheric pressure

These are the Kneser and structural relaxation processes. In systems with small molecular weight m at high temperatures, the Kneser type of relaxation plays the main role [when $\partial(\alpha_p/\omega^2)/\partial T > 0$]. For large m and low temperatures, the behavior of α_p/ω^2 is dominated by structural relaxation [when $\partial(\alpha_p/\omega^2)/\partial T < 0$] [35].

The curves in Fig. 2.8 show how the value of α_p/ω^2 depends on the molecular weight m. It should be noted that there is a pattern in the behavior of $\alpha_p/\omega^2 = f(m)$ for aromatics and paraffins.

In accordance with modern phenomenological theory [35, 38], the ultrasonic absorption coefficient for liquid systems is expressed as

$$\alpha_p = \alpha_p^0 + \alpha_p^v = \frac{8\pi^2\omega^2}{3\rho\vartheta_c^3}\eta_0 + 2\frac{\pi^2\omega^2}{\rho\vartheta_c^3}\eta_v , \qquad (2.17)$$

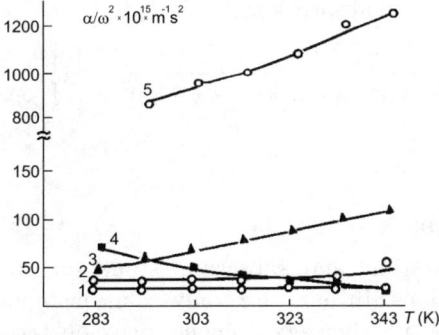

Fig. 2.7. Dependence of α_p/ω^2 on temperature in normal aromatics: (1) n-xylol, (2) ethylbenzene, (3) toluol, (4) tert-butylbenzene, (5) benzene

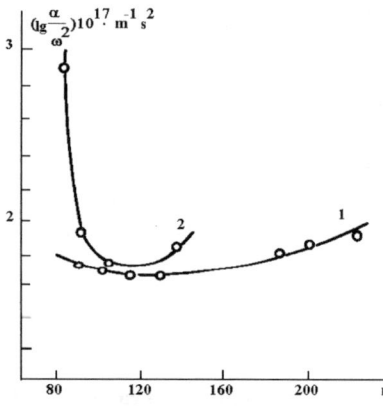

Fig. 2.8. The dependence of α_p/ω^2 on molecular weight for paraffins (1) and aromatics (2)

where α_p^0 is the ultrasonic absorption coefficient due to shear viscosity η_0 (Stocks absorption) and α_p^v is the ultrasonic absorption coefficient determined by bulk viscosity η_v. This equation couples bulk and shear viscosity with the ultrasonic absorption coefficient. It is of great interest from both a scientific and a practical point of view to study the bulk viscosity conditioned directly by dissipation of ultrasound energy due to perturbation of physical or chemical equilibrium in a system while varying pressure or temperature in the wave. This is only a part of the absorbed energy, but it enables one to obtain some kinetic information about fast relaxation processes in liquid systems. Determination of the bulk viscosity may be of practical interest for calculating flow velocities in superfast fluxes or the propagation of shock waves, or again for solving many problems involving η_v in technical acoustics.

Table 2.3 presents the values obtained from (2.17) for all studied temperatures and concentrations of n-heptane–n-nonane and o-xylol–heptane solutions.

Investigations of binary mixtures also bring out the ambiguous character of the concentration dependence of α_p/ω^2. The ultrasonic absorption theory in binary mixtures of Kneser liquids [43] provides the following expression for the concentration dependence of ultrasonic absorption:

$$\frac{\alpha_p}{\omega^2} = \frac{\alpha_{p(B)}/\omega^2}{[\vartheta_{c(A)}/\vartheta_{c(B)}]x_A + x_B} \left[\frac{x_A}{[\alpha_{p(B)}\vartheta_{c(B)}/\alpha_{p(A)}\vartheta_{c(A)}]x_A + x_B} + x_B \right], \tag{2.18}$$

provided that

$$\alpha_{p(A)} \gg \alpha_{p(B)}, \quad \tau_{AA} \gg \tau_{BB}, \quad \tau_{BB} \approx \tau_{AB} \approx \tau_{BA}, \tag{2.19}$$

where τ is the fluctuation relaxation time and subscripts stand for the corresponding intermolecular interactions. In our case, only solutions containing benzene as one of components (i.e., benzene–toluene, benzene–tert-butylbenzene, benzene–hexane, benzene–octane, benzene–hexadecane) satisfy the condition (2.19). Equation (2.18) is applicable only for these solutions.

Table 2.3. Values of η_V [10^{-2} Pa s] and β_S [10^{-10} Pa^{-1}] for different solutions at various temperatures and concentrations and at atmospheric pressure

T [K]	0.0 m.f.		0.2 m.f.		0.4 m.f.		0.6 m.f.		0.8 m.f.		1.0 m.f.	
	η_V	β_S	η_V	β_S	η_V	β_S	η_V	β_S	η_V	β_S	η_V	β_S
n-heptane–n-nonane												
293	0.284	8.84	0.281	9.142	0.264	9.689	0.257	10.006	0.236	10.553	0.232	10.830
303	0.275	9.543	0.254	10.043	0.252	10.387	0.243	10.747	0.216	11.447	0.207	11.875
310	0.250	10.418	0.239	10.954	0.231	11.359	0.216	11.872	0.202	12.242	0.185	12.987
320	0.234	11.182	0.220	11.803	0.216	12.121	0.0197	12.936	0.175	13.602	0.167	14.149
330	0.211	12.398	0.206	12.854	0.195	13.332	0.184	13.873	0.165	14.591	0.152	15.429
340	0.191	13.552	0.182	14.108	0.176	14.545	0.164	15.144	0.150	15.963	0.140	16.805
o-xylol–n-heptane												
283	0.254	9.987	0.322	8.704	0.370	8.043	0.413	7.542	0.573	6.538	0.736	5.772
293	0.232	10.830	0.287	9.542	0.343	8.726	0.398	8.079	0.555	7.032	0.714	6.186
303	0.208	11.875	0.251	10.571	0.307	9.649	0.385	8.694	0.523	7.654	0.699	6.581
310	0.185	12.987	0.227	11.488	0.289	10.349	0.351	9.462	0.518	8.074	0.658	7.135
320	0.167	14.149	0.194	12.852	0.254	11.531	0.348	10.049	0.484	8.730	0.653	7.598
330	0.157	15.429	0.172	14.236	0.226	12.862	0.317	11.152	0.472	9.415	0.627	8.285
340	0.135	16.998	0.150	16.069	0.222	13.704	0.311	11.730	0.446	10.527	0.639	8.884
350	–	18.782	–	18.424	–	15.176	–	12.732	–	11.515	–	9.664

As can be seen from Table 2.4, the experimental dependence of the ultrasonic absorption coefficient is in qualitative agreement with the calculated dependence. This confirms the correctness of the assumption that the Kneser mechanism underlies ultrasonic absorption in the systems considered. For any solution that does not satisfy the condition (2.19), the character of the

Table 2.4. Experimental (I) and calculated (II) values of α_p/ω^2 $\left[10^{-15}\ \text{s}^2/\text{m}\right]$ for various solutions. Calculated values are found using (2.18)

X [m.f.]	293 K		303 K		313 K		323 K		333 K		343 K	
	I	II	I	II	I	II	I	II	I	II	I	II
Benzene–tert-butylbenzene												
0.2	76	83	66	77	57	75	55	73	48	71	32	67
0.4	76	83	66	77	57	75	55	73	48	71	32	67
0.6	140	139	126	130	115	128	110	126	105	123	100	117
0.8	210	252	205	244	186	244	196	243	214	241	224	233
Benzene–n-hexane												
0.2	60	56	61	58	63	60	69	63	72	65	–	–
0.4	83	64	85	67	88	69	93	72	97	74	–	–
0.6	125	87	132	92	146	95	150	100	160	103	–	–
0.8	230	167	255	176	295	181	310	191	340	198	–	–
Benzene–n-octane												
0.2	55	51	55	52	56	53	57	54	59	55	64	57
0.4	74	59	75	61	77	62	78	64	96	65	98	66
0.6	100	84	105	86	117	88	124	90	130	92	140	95
0.8	217	161	221	167	228	171	237	177	254	183	304	187

concentration dependence of α_p/ω^2 should be interpreted separately by taking into account individual peculiarities and properties of the components. An approximate calculation of α_p/ω^2 can be carried out on the basis of the additivity rule with accuracy not worse than ±15–20%.

This means that the linear approximation in the analysis of properties of ideal solutions is broken, and so are the solution additivity and athermity that are a consequence of this linear approximation. Actually, the above experimental results demonstrate nonlinearity of the adiabatic compressibility isotherms. It evidently depends on the type of admixtures in the solution, i.e., even in simple liquid solutions the processes connected with correlation of internal parameters of a system play an important role. As a result, the former division of a mixture of simple liquids into crystallites, or domains, whose behavior is considered to be independent of each other, becomes unacceptable. It seems that one can talk about the appearance of 'prototypes' for a micro-heterogeneous state.

Signs of the self-organization process appearing even in simple systems can be of key importance in creating stable micro-heterogeneous systems with microemulsion properties. A prerequisite for this is the nonlinear dynamics of the system itself.

Therefore, for further explanation of the state-forming mechanisms in emulsions and microemulsions, it is necessary to carry out research on non-linear dynamics of liquid solutions near critical points where correlation processes at large scales are in opposition to the fluctuation dynamics. It is fluctuations that lead to phase-separation of solutions. Thus, one may arrive at microemulsion production by studying ideal solutions through investigation of phase-separation mechanisms and critical dynamics.

3. Phase-Separating Solutions

Most liquid mixtures are far from ideal. Significant differences in the interaction energies of heterogeneous and homogeneous molecules are characteristic of such solutions. If the interaction energy of homogeneous molecules prevails over that of heterogeneous components, then this solution has higher concentration fluctuations. Additivity of the properties is then broken since the more complex properties of near-order configurations must be considered. A significant difference in the properties of heterogeneous molecules and interaction energies of heterogeneous and homogeneous components causes fluctuation regions to grow in size and this leads to phase-separation of the solution. In order to determine properties of this type of system, fluctuation regions must be taken into account.

After the van der Waals equation had been derived, a number of attempts were made to construct a unified equation for the gas and liquid state. The impossibility of obtaining such an equation is due to the existence of individual multiparticle interactions. Nevertheless, systems near critical points possess some common regularities that may be revealed within the framework of the scale invariance theory. The universal character of regularities related to phase transitions allows one to deduce the isomorphism of various critical phenomena of different nature [14, 44–46]. It can be proved that properties of systems near critical points can be described by rather simple exponential functions, where the indices, the so-called critical indices, have the same value for all systems belonging to the same universality class. To determine properties of systems near critical points, critical indices and critical amplitudes must be known.

3.1 Hydrogen Bonds in Solutions with Lower Phase-Separation Critical Point

The reports [47–51] have shown that the appearance of a lower phase-separation critical point is caused by the existence of strongly oriented bonds. These bonds are hydrogen bonds in the case of alcohol–water solutions. When the temperature and the system entropy become greater, this in turn leads to destruction of hydrogen molecular bonds and to further phase separation of the solution. In order to understand these critical phenomena, we must apply an exactly solvable model of a decorated lattice [49] for solutions with hydrogen bonds. This model has been further developed in [48]. One consequence is that an upper critical point of phase separation in systems with hydrogen bonds may be located beneath a lower critical point of a closed phase-separation region. This assumption has been confirmed by experiments on 2-butanol–water, and 2-butanol–water + propanol solutions (see Fig. 3.1)

If directional bonds reside not only between different but also between identical molecules, directional and non-directional interactions between pairs of identical molecules can be stronger than those between different ones. Then a liquid will tend to phase-separate at low temperatures.

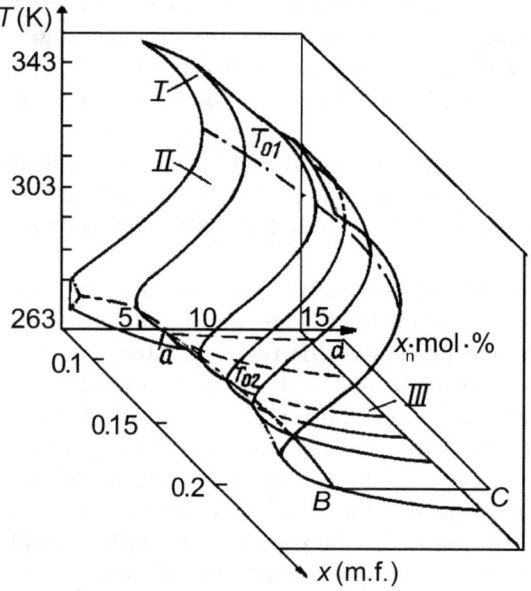

Fig. 3.1. Phase diagram of a 2-butanol–water + propanol solution

However, if the difference between the total interaction energy of identical molecules and different molecular pairs is small in comparison with the energies of directional bonds, a liquid can still be expected to have hydrogen bonds above the upper solution point. At higher temperatures, destruction of hydrogen bonds causes the phases once again to separate. This model reflects those particular systems in which changes in the energy parameters cause the lower critical point of the low temperature region to merge with the upper one to form a double critical point (DCP). The exponent characterizing the form of the coexistence curve is renormalized according to $\beta^* = 2\beta$, where $\beta = 0.33$ is the critical exponent of the coexistence curve for an isolated critical point.

By decreasing the inner region of the closed curve, it is possible to obtain an ordinary double critical point. For certain energy values in this system, the DCP and the upper (low temperature) critical point may merge, thus leading to a so-called critical inflection point. At this point, the critical index is renormalized according to $\beta^* = 3\beta$.

3.2 Phase Diagrams of Phase-Separating Solutions. Order Parameter

Coexistence diagrams of phase-separating solutions are rather varied. At constant pressure, most liquid solutions (e.g., the methanol–heptane system) have the upper critical point of phase separation (see Fig. 3.2), whilst others (e.g., water–triethylamine solution) have the lower one.

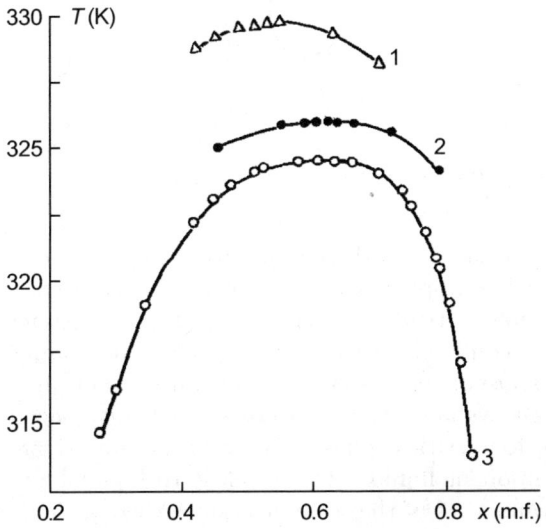

Fig. 3.2. Coexistence curves of the methanol–heptane mixture: (1) [49], (2) [50], (3) [51]

There are also binary mixtures with two and even three critical phase-separation points, e.g., water–2-butanol solution [52, 53]. Apparently, the number of these points may be even greater. In binary solutions, critical point lines already exist in coordinates of pressure, component concentrations, and temperature. Depending on the pressure, these points can merge with subsequent formation of so-called hypercritical points. The simplest example of this merging process is a double critical point arising as a result of drawing together the phase-separation region with the upper and lower critical points.

Such a point is found in a water–2-butanol solution (Fig. 3.1) where, depending on the pressure, there can be two double critical phase-separation points. In accordance with the Gibbs rule, the number of simultaneously coexisting liquid phases of a binary solution cannot exceed two.

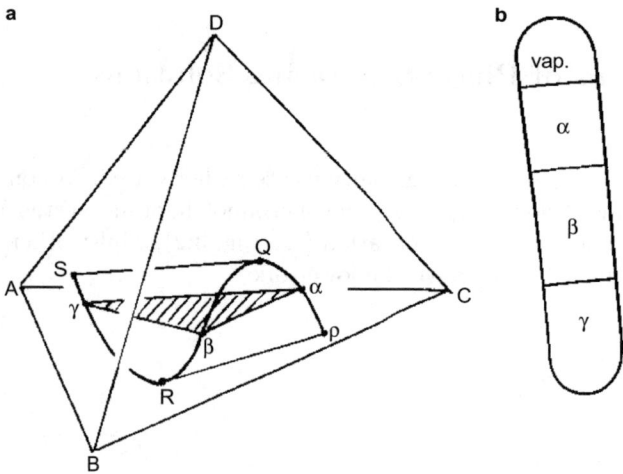

Fig. 3.3. a,b. Phase diagram of a three-component system

In multicomponent liquid solutions (with three or more components), critical points of even higher order may appear with three or more phases coexisting simultaneously, e.g., a triple critical point (Fig. 3.3). The coexistence region of three-liquid phases is confined by the curve PQRS. The triangle $\alpha\beta\gamma$ corresponds to a certain state of the system at a certain ratio of phase components. A water–ethyl–salt system which depends on the salt concentration and pressure can have four critical phase-separation points [57, 58]. If we are to describe multicomponent liquid solutions close to liquid–liquid phase transitions within the framework of the scale invariance theory, then a proper choice of the order parameter must be made.

For one- and two-component systems, such a parameter may be the difference in concentrations of identical components in coexisting phases, and there is no room for ambiguity in the choice. In three-component mixtures, isomorphism is maintained at constant chemical potential of the third component [14, 59]. This condition is rather difficult to realize in a real experiment. When studying multicomponent solutions, the problem of how to make an unambiguous choice of order parameter remains unsolved.

3.3 Phase Diagrams of Binary Solutions with One Critical Point

The phase diagram of a binary phase-separating solution (binodal curve) in the framework of the scale invariance theory can be described by the relation

$$x - x_c = \pm A t^\beta + F(T), \qquad (3.1)$$

where x is the concentration of any component of the upper $(+)$ or lower $(-)$ phase when the solution separates, x_c is the critical concentration, $t = (T_c - T)/T_c$ is the reduced temperature, T_c is the critical temperature of phase separation, T is the solution temperature, A is a constant, β is the critical index of the order parameter, and $F(T)$ is a function that accounts for the asymmetry of the binodal curve. Thus, (3.1) gives us the difference of concentrations of coexisting phases which can be chosen as an order parameter

$$\phi = x_1 - x_2 = 2At^\beta .$$

Solid lines in Fig. 3.2 present the processed experimental phase diagram of the methanol–heptane solution.

Values of the critical index of the order parameter obtained by experiment for a wide range of binary solutions (i.e., organic solutions, solutions with strong interactions, conducting solutions, metal solutions, metal–ammonium solutions with free electrons, etc.) ranged from 0.32 to 0.35. Values obtained by the three-dimensional Ising model [60] and in other ways (summing rows, solving the renormalization group equations, ϵ expansions) turned out to be the same.

3.4 Binary and Ternary Solutions with Closed Phase-Separation Region

Phase diagrams for most of the systems given below have been obtained by visual observation of meniscus formation in a closed vessel put into a thermostat. The phase diagrams of propanol–water + NaCl, butanol–water + NaCl, 2-butanol–water + pressure, 2-butanol–water + propanol, methylethylketone–water + acetone, guaiacol–glycerine + water, β-picoline–water + heavy water systems have undergone the most detailed study.

Figure 3.4 presents the x_1, x_n, T phase diagram of the propanol–water + NaCl solution, where x_1 [m.f.] (mole fraction) is the concentration of the first component in water, x_n [m.f.] is the admixture concentration in the solution, and T is the temperature. The lines of upper and lower critical phase-separation points are given in the (x_0, T) coordinate plane and, in total, form a closed surface separating the closed ordered region (a phase-separation gap) from the single phase disordered region. A characteristic

28   3. Phase-Separating Solutions

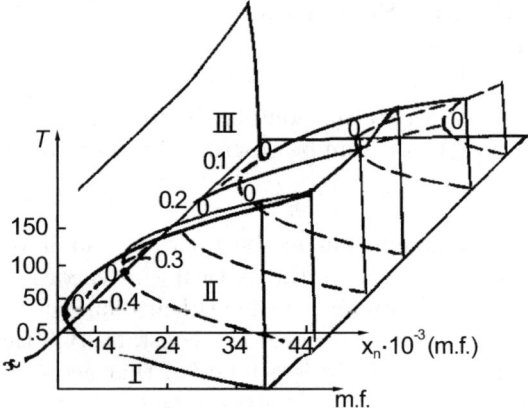

Fig. 3.4. Phase diagram of the propanol–water + NaCl system

peculiarity of the diagrams is the double point lines (dashed lines) dividing the phase-separation surface into two parts with upper and lower phase-separation temperatures T_c.

The location of a double critical point on the double point line, in which the upper and lower critical points merge, is unknown beforehand, and as a rule it does not coincide with the top of the surface of phase-separation points. Determination of the DCP parameters (x_{01}, x_{0n}, T_0) is therefore an independent experimental problem. To solve it, in [59], we used the condition of DCP coincidence with the intersection of the fold points and double point lines. In addition to the direct method for finding the DCP, it was suggested that a decision could be reached through the dependence of double point lines on the concentrations of basic components.

As can be seen from Fig. 3.5, the DCP location on the curve $T(x)$ always corresponds to the section of a minimal slope, i.e., it coincides with the minimum point of dT/dx, where $d^2T/dx^2 = 0$. Note that this condition, being a consequence of symmetrization of the phase diagram in the vicinity of the DCP, was observed for all the systems studied. Consequently, it may be used as an accurate and rather simple method for evaluating all DCP parameters and it is not inferior to competing methods.

The admixture plays a double role in the above-mentioned systems with DCP. Increasing its amount may induce either convergence (in the butanol–water + NaCl, 2-butanol–water + propanol solution) or separation (in the propanol–water + NaCl, guaiacol–glycerine + water solution) of the upper T_B and lower T_H temperatures. Note that an increase in pressure P in these systems causes convergence of T_B and T_H to a DCP. Beyond the DCP, a solution can exist only in a homogeneous state (absolute solubility region).

As mentioned, the DCP location in the systems under study does not coincide with the top of the surface in the coordinates x_1, x_n, T. This is due to the fact that x_{01} does not correspond to the extremum point on the equi-

3.4 Binary and Ternary Solutions with Closed Phase-Separation Region

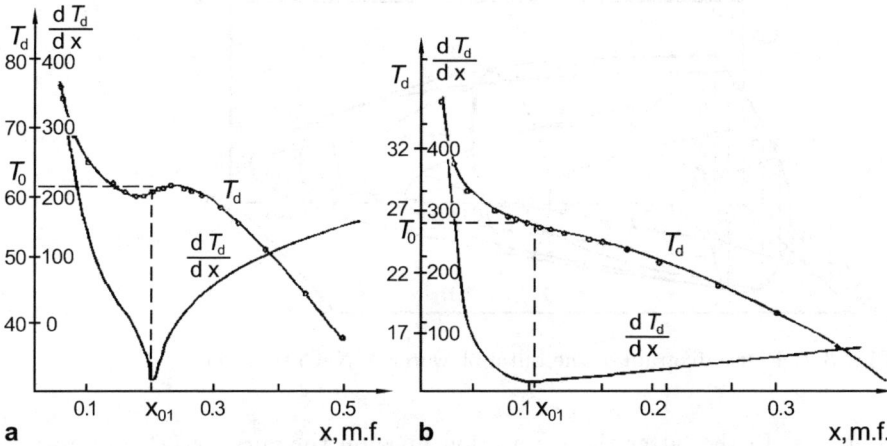

Fig. 3.5. Temperature dependence of double points T_D on the concentration of the main components for methylethylketone–water + 1,4-dioxane (**a**) and butanol–water + HCl solutions (**b**)

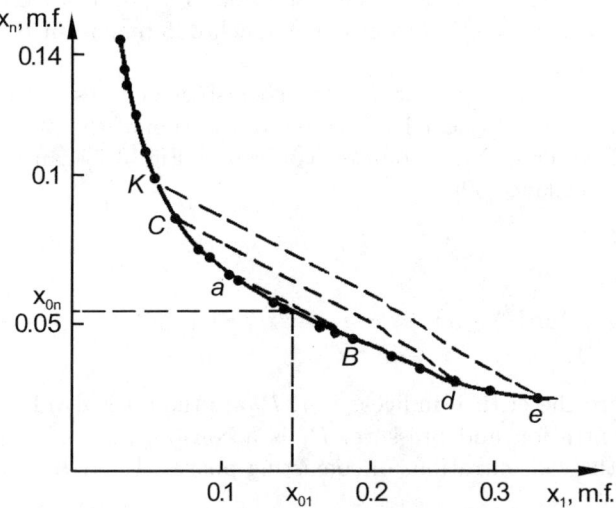

Fig. 3.6. Phase diagram of the propanol–water + NaCl solution at 318.1 K

librium curve $x_1(x_n)$ (Fig. 3.6). Such an asymmetry is a consequence of the non-symmetrical distribution of the admixture concentration x_n between the solution phases upon phase separation. For this very reason, the connection lines between points ab, cd, kl on the curves $x_1(x_n)$ are not parallel to the x axis. This is a significant point in understanding why the phase diagrams of solutions with admixtures always differ so much from those of binary DCP

30   3. Phase-Separating Solutions

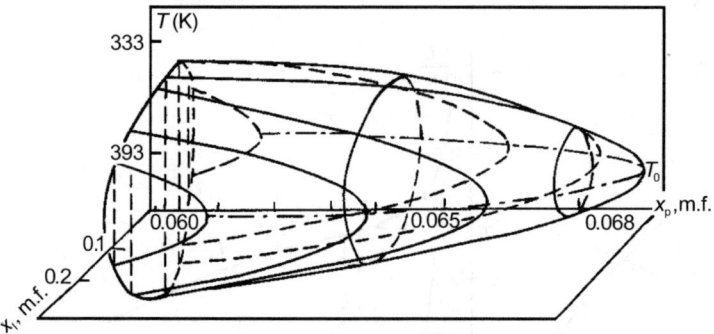

Fig. 3.7. Phase diagram of the butanol–water + NaCl solution

solutions. In the latter the connection lines on the curve $x_1(P)$ are always parallel to the x axis.

A solution with admixtures simulates a quasi-binary solution when the admixture solubility of both components is identical. This seldom happens. Note, however, that the butanol–water + NaCl solution (Fig. 3.7) is close to quasi-binary because x_{01} for this solution is not much different from the maximum of the curve $x_1(x_n)$.

In ideal quasi-binary solutions, the form of the coexistence curves in the coordinates (x_1, T), (x_n, T), (P, T) near DCP (the cross-section of the phase diagram of the butanol–water + NaCl solution is given in Fig. 3.7) can be approximated by the power laws [59]:

$$\begin{aligned} & \mid x_1 - x_2 \mid \sim \mid T_c - T \mid^{\beta_T} , \\ & \mid x_1 - x_2 \mid \sim \mid x_{c\Pi} - x_{0\Pi} \mid^{\beta_\Pi} , \\ & \mid T_c - T_0 \mid = B_1 \mid x_{c\Pi} - x_{0\Pi} \mid^{\beta_0} , \\ & \mid T_c - T_0 \mid = B_1 \mid P_c - P_0 \mid^{\beta_P} . \end{aligned} \quad (3.2)$$

where β_T, β_Π, β_0, β_P are the critical indices, $x_{c\Pi}$, P_c are the critical values of the admixture concentration and pressure, B_1 is a constant, $x_1 - x_2$ is the difference between the concentrations of coexisting phases A and B (the order parameter).

This results in the following relationship between critical indices of the order parameter at DCP:

$$\beta_\Pi = \beta_0 \beta_T . \quad (3.3)$$

The results of [59, 61, 62] show that, within the range of experimental error, the index β_0 is equal to 0.5 for all systems without exception.

According to (3.3), this means doubling the index β_T, which agrees with theoretical predictions [14, 59, 63]. Obviously, employing (3.3) to determine the index β_T is difficult. Actually, being a parabolic equation, it only holds in a very small neighborhood of T_B and T_H and cannot describe the (T, x)

3.4 Binary and Ternary Solutions with Closed Phase-Separation Region

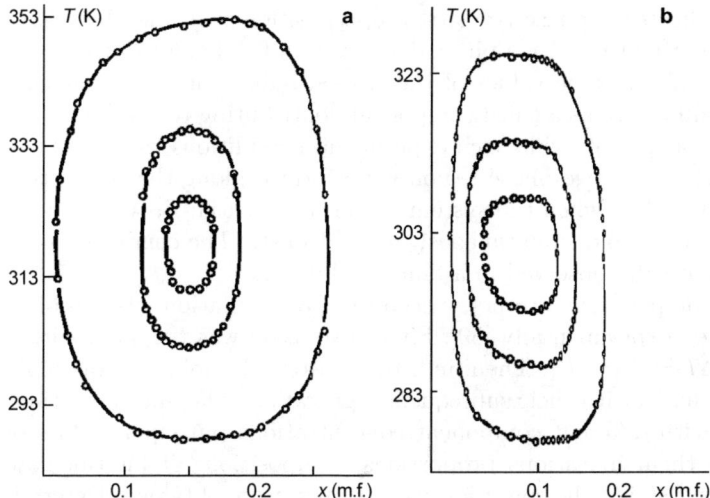

Fig. 3.8. (T,x) phase coexistence diagrams of solutions in the quasi-binary approximation for propanol–water + NaCl (**a**) and butanol–water + NaCl (**b**) systems

diagram as a whole close to DCP, since the latter has a closed, almost elliptical form (see Fig. 3.8a and b)

It is natural to assume that the observed form of the equilibrium curve $T(x_1)$ is stipulated by the closeness of the critical temperatures T_B and T_H, which correspond to two phase transitions and define the limits of system stability. Therefore, the form of the boundary curve turns out to be a function of two variables, depending on the closeness of the system temperature T to T_B and T_H. In this case, the boundary curve in the (T, x_1) plane can be described by

$$|x_1 - x_2| = A_1 |T_B - T|^{\beta_{T1}} |T_0 - T_H|^{\beta_{T2}} . \tag{3.4}$$

In the particular case where the system temperature equals T_0 and $\beta_{T1} = \beta_{T2} = \beta_T$, we have

$$A_2 |x_1 - x_2| = |T_B - T_0|^{\beta_{T1}} |T_0 - T_H|^{\beta_{T2}} . \tag{3.5}$$

Using the fact that

$$L|x_1 - x_2| = |x_{c\Pi} - x_{0\Pi}|^{\beta_\Pi} , \tag{3.6}$$

we have

$$|T_c - T_0| = \frac{A_2}{L} |x_{c\Pi} - x_{0\Pi}|^{\beta_\Pi / 2\beta_T} . \tag{3.7}$$

Comparing with (3.2), we obtain $\beta_\Pi = \beta_T$, since $\beta_0 = 0.5$.

As shown by the studies, the form of the boundary curve in the (T, x_1) plane, obtained experimentally for the butanol–water + NaCl solution (see Fig. 3.7), is fairly accurately expressed by (3.4).

Since (3.4) reflects the phase-transition superposition principle, this makes it possible to describe the (T, x_1) phase diagram at a fixed value of the index β_T. However, to obtain true values of the critical indices of a real solution, containing an admixture as a third component contributing to the boundary curve, it is necessary to provide such experimental conditions that a change in the composition in a two-phase region is performed along the connection line (e.g., Fig. 3.6), i.e., when the system chemical potential $\mu_T = $ Const. at constant temperature and simultaneously $x_{II} \neq $ Const. (The condition $\mu_T = $ Const. is automatically observed in a binary solution.)

Since such compositions are not susceptible to calculation, the problem has been resolved experimentally [59]. First, a solution was prepared with a given value of $\Delta T = T_B - T_H$. Then, in a thermostat, the solution was separated into two phases with their subsequent segregation. The initial solutions were obtained with different component concentrations but equal values of μ_T. By mixing them in various proportions, compositions of intermediate concentrations satisfying the condition $\mu_T = $ Const. were obtained. Determination of the component concentration in the original solutions, needed to calculate the intermediate compositions, was performed by quantitative physico-chemical analysis. The diagrams of composition versus phase-separation temperature are shown in Fig. 3.8.

Table 3.1 displays critical indices of the order parameters β_{T1}, β_{T2} and β_0, as well as the coefficient B_1 found by approximation of (3.2) and (3.4). The approximation interval of t ranged over 10^{-3}–10^{-1}. In all systems investigated it follows from Table 3.1 that the index β_T lies in the range 0.34–0.38 and close to the value of 0.36 for a pure binary butanol–water solution. Quantitative values of the indices are kept either close to DCP or far from it, within the

Table 3.1. Critical indices of various solutions

Solution	B_1/X	$\beta_{T1}/\Delta T$	$\beta_{T2}/\Delta T$	B_0/X
Propanol–	–	0.40/0.5	0.34/0.5	0.51/0.5
water + NaCl	–	0.38/0.3	0.37/0.3	0.51/0.1
	–	0.38/0.1	0.37/0.1	0.51/0.2
Butanol–	448/0.05	0.40/0.5	0.33/0.5	0.51/0.05
water + NaCl	455/0.07	0.38/0.3	0.35/0.3	0.49/0.15
	455/0.10	0.37/0.1	0.36/0.1	0.50/0.10
	446/0.20	–	–	0.50/0.20
Methylethylketone–	372/0.2	–	–	0.50/0.10
water + propanol	–	–	–	0.49/0.15
	–	–	–	0.50/0.20
	–	–	–	0.50/0.30

3.4 Binary and Ternary Solutions with Closed Phase-Separation Region

limits of experimental error. It is characteristic that the value of β_T is always slightly higher for the upper critical points than for the lower ones.

The critical indices of an external field β_0 and β_P are supposed to be equal. Despite this, the above method can completely describe the behavior of the order parameter critical indices for a closed phase-separation region. An isomorphous choice of the parameter itself is open to question.

To elucidate this problem, measurements of the temperature and concentration dependencies of the electrical conductivity and IR spectral absorption in single phase and two-phase regions were performed in [64,65]. In the presence of an admixture component and taking into account the constancy of the pressure, (3.4) reduces to

$$\Delta x = A_1 \left[(T - T_0)^2 - \Delta T^2 \right]^\beta , \qquad (3.8)$$

where Δx is the difference between the concentrations of the coexisting phases of the solution components. Note that a proper use of (3.8) is only possible when there is compositional change of the basic components at constant chemical potential of the admixture component. Equations (3.4) and (3.8) have been confirmed in [59] for the cross-sections obtained experimentally along the connection line. Here, constancy of the connection line slope is implicitly assumed when the solution temperature is changed.

Suppose that (3.8) is observed for three- and multicomponent solutions:

$$\begin{aligned} x_i &= \pm A_i[(T - T_0)^2 - \Delta T^2]^\beta + F(T) , \\ x_i &= |x_\Pi - x_{0\Pi}|^{\beta_\Pi} + V(T) , \end{aligned} \qquad (3.9)$$

where x_i is the concentration of the ith component in separated phases, whilst functions $F(T)$ and $V(T)$ take into account the asymmetry of the phase separation curve.

The experimental difficulty in checking (3.8) and (3.9) lies in finding the concentration of each component in the solution. Therefore, in order to process the experimental results, the electrical solution conductivity k and IR absorption α_i were used instead of component concentrations in coexisting phases. Both for the temperature and concentration dependence, the behavior of the electrical conductivity (IR absorption) characterizes changes in the system component concentrations of corresponding phases. This is because there is an unambiguous correspondence between solution component concentrations and electrical conductivity. One may assume that the values of k and α_i are proportional to the order parameter and can describe its behavior within the framework of statistical scaling. Bearing in mind the superposition of phase transitions [59, 64] for k or α_i, we can write

$$k = \pm A[(T_B - T)(T - T_H)]^{\beta_T} + B + CT + DT^2 , \qquad (3.10)$$

where A is the scale factor, B, C, and D are coefficients expressing the asymmetry of the phase-separation curve and location of the direct diameter. Analogously, for changes in the admixture concentration at constant temperature, we have:

$$k = \pm A_1 |x_n - x_{0n}|^{\beta_\Pi} + B_1 + C_1 T + D_1 T^2 . \tag{3.11}$$

Figures 3.9 and 3.10 display the temperature dependencies of the solution conductivity k and its IR absorption α_i in heterogeneous regions at various admixture concentrations. The curves $k(T)$ and $\alpha_i(T)$ form a closed region. The difference in k and α_i for coexisting phases is maximal at $T = T_0$ and tends to zero at the temperatures T_B and T_H.

IR absorption measurements were performed at wavelength $6\,900$ cm^{-1}, corresponding to the maximum of the absorption band of the first fluctuation overtones of hydrogen bonded molecules. The results of processing the measurements by formulas (3.8) and (3.9) are shown in Figs. 3.9 and 3.10 (solid lines).

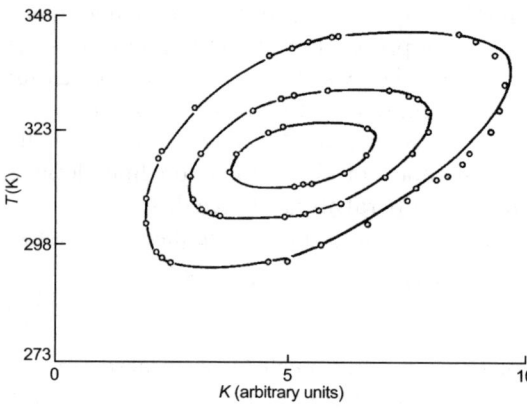

Fig. 3.9. Conductivity of coexisting phases of the propanol–water + NaCl solution

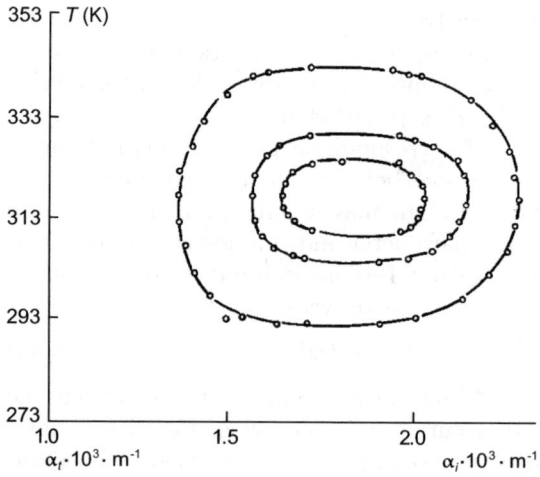

Fig. 3.10. Amplitude of IR absorption in the coexisting phases of the propanol–water + NaCl solution

3.4 Binary and Ternary Solutions with Closed Phase-Separation Region

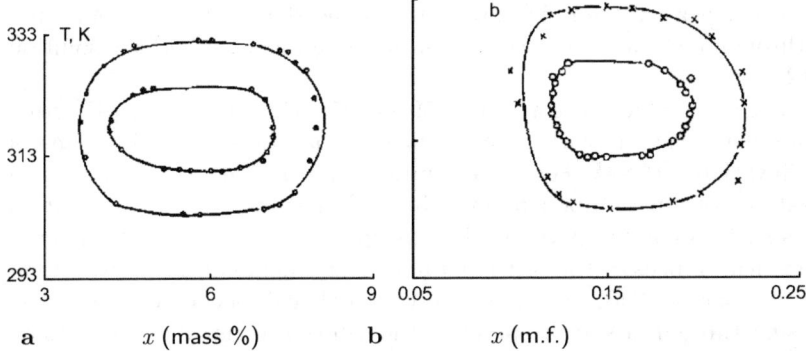

Fig. 3.11. Temperature dependence of the distribution of NaCl in coexisting phases (a) and the main component concentration in the layers (b) for different widths of the phase-separation region

Critical indices of the conductivity ($\beta_T = 0.33\pm0.03$ and $\beta_\Pi = 0.33\pm0.03$) and IR absorption ($\beta_T = 0.33\pm0.05$ and $\beta_\Pi = 0.33\pm0.05$) are in agreement with theoretical values and those obtained from direct measurements [64].

For further confirmation of the adequacy of (3.8), concentrations of each component in coexisting phases were calculated in [64, 65]. Highly accurate data on conductivity helped to restore the distribution of all components in coexisting phases (Fig. 3.11).

The adequacy of the temperature and concentration dependencies of the heterophase state suggests that an isomorphous description of the phase diagram of a multicomponent solution is suitable if the difference in concentrations of any component in the coexisting phases is chosen as order parameter.

Note that (3.8) and (3.9) result in one more interesting conclusion. Let us transform (3.9) into

$$\frac{\Delta x_1}{A_1} = \frac{\Delta x_2}{A_2} = \frac{\Delta x_3}{A_3}, \tag{3.12}$$

where the temperature dependence is absent. The relation (3.12) in its turn implies linear dependence of the coexisting phase concentrations on one an-

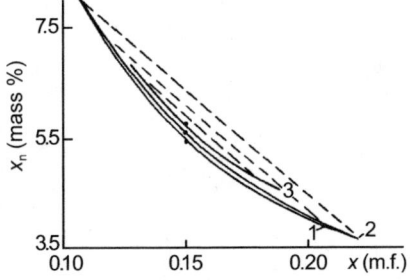

Fig. 3.12. Phase diagrams of the solution at fixed temperatures. *Dashed lines* correspond to connection lines: (1) 313 K, (2) 318 K, (3) 323 K

other. In other words, it implies constancy of the slopes of connection lines passing through a certain figurative point at any temperature of the solution (Fig. 3.12).

It is useful to review the reports [59, 64, 65] when seeking to describe multicomponent systems with a closed phase-separation region in terms of the scale invariance theory. For these solutions, we can take the difference in concentrations of any component including admixtures, as well as solution properties such as conductivity and IR absorption, as an order parameter.

The critical indices of the order parameter are identical for a given temperature, pressure and admixture concentration. The directions of connection lines passing through a specific point of the solution phase diagram remain constant when the temperature is changed.

4. Dynamics of States Close to Critical

4.1 Low Frequency Acoustic Spectroscopy of Weakly Absorbing Liquids

Studies on relaxation processes in condensed matter by molecular acoustics methods have been of consistent interest for decades [33–43, 66–72]. The appearance of new, precise, low-frequency methods [73–77] and the discovery of objects with unusual phase states (e.g., solutions with structural phase transitions and merging critical points, micelle-forming solutions, liquid crystals, etc.) led to intensive investigations in this direction [78–94]. A wide range of possible equilibrium states were revealed, with formation of several macro-heterogeneous regions and merging critical points, as well as thermodynamically stable micro-heterogeneous regions and corresponding critical points of micelle-formation, structural phase transitions, etc.

When employing the impulse method at frequencies lower than 5–10 MHz, difficulties arise from the heavy influence of diffraction on the obtained results. Diffraction formulas for a piston radiator field enable one to make up for inaccuracies in the measurement of acoustic liquid parameters, but only up to a certain point, since the surface amplitude distribution of a real piezo-radiator and a piston radiator field differ, according to the substitution method [75]. When standard and studied liquids with close velocities of ultrasound propagation are separated by a sound-transparent baffle plate, and the distance between the radiator and the sound receiver is fixed, diffraction inaccuracies are decreased by an order of magnitude. Measurement inaccuracies at lower frequencies of the investigated range are high enough to view them as unreliable.

A much smaller amount of liquid with the same number of diffraction errors is needed for measurements by an ultrasound resonator with flat piezo-plates [73, 74, 90, 95]. Trustworthy results obtained by this method are explained by small diffraction losses in the resonator. Piezo-plates of large diameter in the interferometer, and with a small separation between them, enable one to determine the optimal ratio of the resonator length and the piezo-plate diameter, when small amounts of the investigated liquid meet with minor losses. In contrast with the impulse method, experimental checks show that diffraction losses in the resonator with flat piezo-plates are well described by the formulas for an open resonator and can be mirrored by calculation. However, in practice, researchers usually use a calibrated liquid to measure diffraction and other losses.

One achievement of the resonator techniques has been the development of measurement methods based on resonance peaks of the frequencies far from the resonance frequency of the piezo-plates. In this case the oscillation intensity of the piezo-plates is less by an order of magnitude than that of the liquid, and so losses in the piezo-plates are small. Applications of calibrated liquids regard energy losses as being due to dissipation of oscillations in the piezo-plates if the acoustic impedances of these liquids are of the same order, i.e., the difference in densities of investigated and calibrated liquids does not bring in a large measurement error. Since the ultrasound beam of a resonator is usually practically normal to the surface of the piezo-plate, the tangential component of the oscillations arising is small, and the difference in shear viscosities of the investigated and calibrated liquids makes an insignificant contribution to experimental results, comparable with the measurement errors.

The simplicity of the measurement method and availability of standard experimental apparatus determines the wide application of the resonator method in various acoustic investigations at frequencies in the range 140 kHz–30 MHz. In [96, 97], a significant decrease in the diffraction losses in a resonator at frequencies below 0.5 MHz was reached by applying an excess pressure of about 1 atm inside a resonator with round, flat piezo-plates. Under this pressure, the flat plates became concave, resulting in a concentration of the ultrasound beam at the center of the piezo-plates and its isolation from the side surfaces. This considerably reduced diffraction losses in the resonator and halved the low-frequency limit of the working range. Analogous results were obtained in [74, 97, 98] by employing concave piezo-converters, or reflectors. However, it was impossible to reduce the intrinsic losses of the resonator. Specifically designed experiments [75] have demonstrated that the observed losses cannot be explained solely by beam diffraction divergence, but are caused by oscillations of the piezo-plates.

The most reliable methods of acoustic measurement are those that reduce the intrinsic losses of resonators to a minimum by their optimized shapes. In [99], the authors describe a resonator with fluorine-layer films which adopts a concave shape when liquid is placed between the films. Excitation of the resonator is achieved with the help of a piezo-ceramic element joined to the films by a liquid drop and excited at a non-resonant frequency. The resonator may be used in measurements of aggressive liquids and has a lower working frequency of 20 kHz. The best are resonators with concave piezo-lenses [96], whose intrinsic oscillation quality is higher by an order of magnitude ($Q = 10^6$) than the typically used resonators with flat, parallel plates. These resonators provide a means to research ultrasonic absorption in weakly absorbing liquids that was previously quite inaccessible. In resonators with concave piezo-lenses, oscillations of the points on their inner surface become equiphase, losses from the piezo-plates decrease considerably, and their oscillation quality increases by an order of magnitude. As shown in [96], radiation energy losses from the resonator piezo-lenses into the air are significantly less than the total parasite losses because piezo-lenses serve as a quarter-wave reflecting layer.

4.2 Acoustic Spectroscopy of Critical Solutions with Low Sound Absorption

In the middle of the 1950s, experimental accuracy increased and it became obvious that a correct description of the experiment in the vicinity of critical phase-separation points should embrace order parameter fluctuations, as their amplitude in the volume of the correlation radius is within the limits of average concentrations. Experimental investigations on acoustic properties of phase-separating solutions [73–76] carried out in the 1960–70s revealed that many solutions with critical phase-separation points had an anomalous growth of acoustic absorption with a wide spectrum of relaxation times and high dispersion of the ultrasound velocity. It was noticed that, when approaching critical points, acoustic spectra rapidly shifted to the low frequency region. Note that critical ultrasonic absorption does not increase in a number of systems (e.g., methanol–n-heptane solution) despite strong optical (critical opalescence) and thermophysical (heat capacity peak) anomalies close to the critical point. By making use of the precise acoustic techniques [55] we have developed, the above solution was found to have anomalies in critical sound absorption at much lower frequencies, with sound absorption coefficients 30–50 times less than those in systems of a similar kind.

Studies on acoustic anomalies in the vicinity of critical points of binary solutions show that a correct choice of the amplitude of critical ultrasonic absorption provides trustworthy acoustic data within the framework of the scaling theory. Calculated amplitudes of critical ultrasonic absorption may sometimes diverge from experimental values by more than an order of magnitude. Moreover, in many cases, the said calculations cannot be performed at all. It is necessary to know some experimentally determined parameters, such as the heat capacity [55], short-range correlation radius, volume extension coefficients, etc.

There are two main approaches to the description of the critical absorption. The first was developed by K. Kavasaki [100], who supposed that the acoustic absorption mechanism in the critical region is connected with the volume viscosity relaxation, whose dependence on ultrasound frequency and relaxation time is determined by

$$\eta_v(\omega, \tau_c) = \omega^{-1}\Im(\omega^*), \tag{4.1}$$

4. Dynamics of States Close to Critical

where $\omega^* = 2\omega\tau_c$, and $\Im(\omega^*)$ is some universal function. Note that the main task in the development of this theory is finding the functional form of $\Im(\omega^*)$. The sound absorption coefficient can be calculated using the well-known formula:

$$\alpha_p = \frac{\omega^2}{2\rho\vartheta^3}\left[\frac{4\eta}{3} + \eta_v + k(C_V^{-1} + C_P^{-1})\right], \qquad (4.2)$$

where k is the conductivity coefficient.

Another approach suggested by R.A. Ferrel and J.K. Bhattacharjee (FB) [101] is based on the mechanism caused by heat capacity relaxation. Analysis within the scope of scaling theories [102] has shown that both mechanisms provide close analytical dependencies of critical absorption and its amplitude.

Sticking to the general concept in further considerations, we shall discuss the scaling theory of the heat capacity relaxation in some detail. The critical solution concentration is assumed to be implicitly connected with changes in pressure. Therefore, two independent variables, viz., pressure P and temperature T, are satisfactory. The adiabatic compressibility can be written [95]

$$\beta_S = \beta_B - \frac{g^2}{T_c\rho C_P}, \qquad (4.3)$$

where β_B is a non-critical part of the compressibility. Viewed within this theory, the ratio of ultrasonic absorption coefficients obtained through ϵ expansion [103] is determined by the expression

$$\frac{\alpha_p}{\alpha_{\max}} = \frac{2}{\pi}(1+p)\omega^* I(\omega^*), \qquad (4.4)$$

where

$$I(\omega^*) = \int_0^\infty \frac{U(1+U)^p dU}{(1+U)^2[\omega^* + U^2(1+U)^{2p}]},$$

$\alpha_{\max}$ is the ultrasonic absorption coefficient at critical temperature T_c, U is the integration variable, the parameter $p = 1$ as $\epsilon \to 0$, $p = 1/2$ as $\epsilon \to 1$.

The frequency dependence of the maximal absorption at $T = T_c$ has the form [102]

$$\left(\frac{\alpha_p}{\omega^2}\right)_{\max} = S\omega_c^{-(1+\overline{\alpha})} + \frac{\alpha_f}{\omega^2}, \qquad (4.5)$$

where the critical absorption amplitude is

$$S = \frac{\pi^2 \overline{\alpha} C_r g^2 \vartheta_c^2}{2\phi\nu T_c C_P^2}\left[\frac{\omega_0}{2\pi}\right]^{\overline{\alpha}\phi\nu}, \qquad (4.6)$$

with

$$\omega_c = \omega_0 t^{\phi\nu} = \frac{k_B T_c}{3\pi\eta_0 r_0^3}t^{\phi\nu}, \quad C_P = C_r t^{-\overline{\alpha}} + C_f,$$

and g a dimensionless connecting constant equal to

4.2 Acoustic Spectroscopy of Critical Solutions with Low Sound Absorption   41

$$g = \rho C_P \left[\frac{dT_c}{dP} - \frac{T_c \Theta}{\rho C_P} \right] . \tag{4.7}$$

In equations (4.5)–(4.7), $\bar{\alpha}$, ϕ, and ν are critical indices of the heat capacity, shear viscosity and correlation radius, respectively, C_r and C_f are regular and non-critical parts of the heat capacity, α_f is the background (non-critical) absorption coefficient, and Θ is the volume extension coefficient.

By comparing (4.6) and (4.3), we find that the first term on the right-hand side of (4.6) is proportional to the singular part of the adiabatic compressibility:

$$S \propto \frac{C_r g^2}{T_c C_P^2} = \frac{C_r}{C_P} \rho \beta_S . \tag{4.8}$$

Investigation of the phase diagrams of methanol–n-heptane and methanol–cyclohexane solutions showed that the critical compositions are 0.616 and 0.515 m.f. of methanol, respectively, and critical temperatures lie in the neighborhood of 325.4 K and 319.45 K.

Figure 4.1 gives the results of ultrasonic absorption investigations on a methanol–n-heptane solution in the vicinity of the critical points $(T - T_c) \to 0$, showing α_p/ω^2 as a function of $\omega^{-1.06}$. The results of fitting by (4.5) (solid line) are in a good agreement with the experiment. Note that the magnitudes of absorption of both solutions are close and differ by a factor of 30 from those obtained experimentally by other authors [77].

Figure 4.2 shows the frequency dependence of absorption α_p/ω^2 for a methanol–cyclohexane solution for various degrees of approach towards the

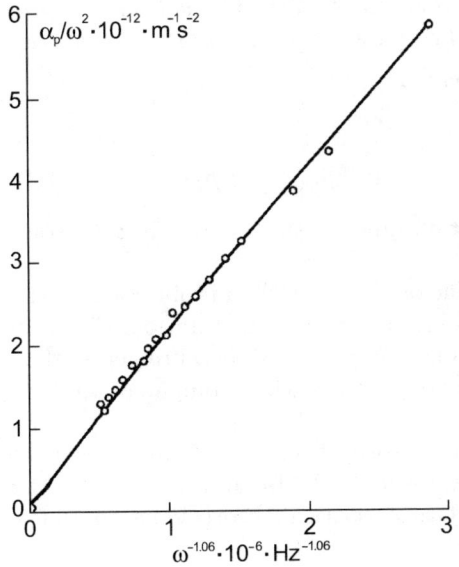

Fig. 4.1. Dependence of ultrasonic absorption on the frequency of a methanol–n-heptane solution in the immediate vicinity of the critical point

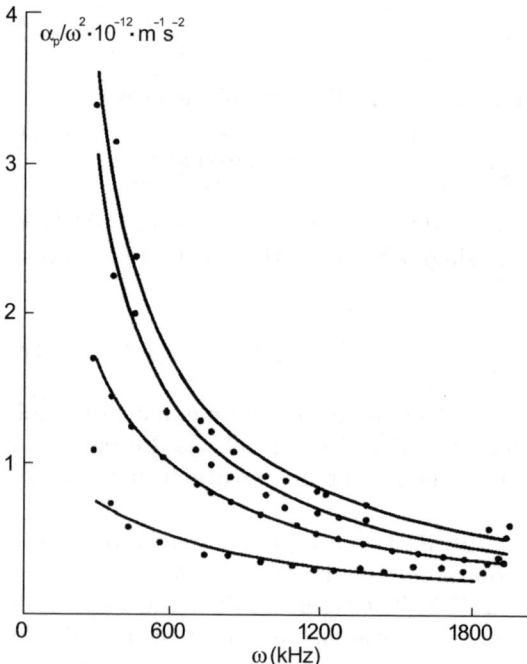

Fig. 4.2. Frequency dependence curves for sound absorption in a methanol–cyclohexane solution at various degrees of approach to the critical phase-separation temperature

critical phase-separation temperature. Solid lines represent the results of processing experimental values by (4.5). The parameter values are obtained as

$$S = 2.18 \times 10^{-6} c^{0.94}/\text{m}, \quad r_0 = 2.67 \times 10^{-10} \text{m}, \quad g = 0.22,$$

and

$$S = 2.1 \times 10^{-6} c^{0.94}/\text{m}, \quad r_0 = 3.84 \times 10^{-10} \text{m}, \quad g = 0.24,$$

respectively, for the methanol–n-heptane and methanol–cyclohexane solutions.

From acoustic measurements, a value of 2.67×10^{-10} m is obtained for the short-range correlation radius for the methanol–n-heptane solution. This is in good agreement with the value determined from optical measurements [102]. The same can be said for the parameters g and S when compared with the values calculated by (4.5).

The regular parts of the solution absorption α_p/ω^2 are equal to 44.3×10^{-15} s^2/m and 73×10^{-15} s^2/m, respectively. Below are presented the values of the constants needed to calculate the critical absorption amplitudes.

- Methanol–heptane system:

  $T_c = 325.4$ K, $\overline{\alpha} = 0.11$ [14],
  $C_S = 140$ J/g, $C_r = 2262$ J/g [103],
  $C_P = 2530$ J/g, $dT_c/dP = 4.2 \times 10^{-7}$ [104],
  $V_{c0} = 983.9$, $\eta_0 = 0.39$,
  $\alpha_V = 1.46 \times 10^{-3}$, $r_0 = 2.6 \times 10^{-10}$ [102],
  $z = 3.06$ [77].

- Methanol–cyclohexane system:

  $T_c = 319.45$ K,
  $C_S = 75.4$ J/g, $C_r = 2310$ J/g [102],
  $C_P = 2460$ J/g, $dT_c/dP = 3.4 \times 10^{-7}$ [104],
  $V_{c0} = 1087$ /s, $\eta_0 = 0.39$ g/s,
  $\alpha_V = 1.28 \times 10^{-3}$, $r_0 = 3.24 \times 10^{-10}$ [102].

Thus, precise acoustic methods can be employed to reveal and study acoustic relaxation in binary methanol–heptane and methanol–cyclohexane solutions.

4.3 Acoustic Perturbation and Correlation Radius of Fluctuations in the Vicinity of a Critical Point

A critical point (CP) is characterized by the appearance of long-range ordering in the system, i.e., by an anomalous growth of the correlation radius. However, numerous experimental results show that the magnitude of the average correlation radius is finite. Since the accuracy of experimental measurements is at best about 10^{-3}, experimental results are obviously not obtained at the critical point itself, but in its vicinity. Hence, extrapolation of the system parameters may not coincide with their real values. Consequently, theoretical analysis of experimental results should take into account the deviation from CP.

A finite correlation radius shows that the system under investigation is characterized by a probability distribution that does not correspond to thermodynamical equilibrium, i.e., in the system there are fluctuations affecting its order parameter. The liquid may be considered as a heterogeneous system consisting of domains with opposite signs of the order parameters. Under these conditions, the influence of an ultrasonic perturbation on a system in the vicinity of a critical point changes the space fluctuations of the order parameter, on account of the interaction between different regions of the system with an external field ultrasound wave. Thus, for the fluctuating value of the order parameter φ and, conjugate with it, the ultrasound field h, an additional term in the system Hamiltonian takes the form:

$$\delta H = h \int \varphi(x) dx .$$

The appearance of such an extra term would affect the establishment of equilibrium in the system. In this case a change in the order parameter is described by the Ginzburg–Landau equation:

$$\dot{\varphi} = a\varphi - b\varphi^3 + g\Delta\varphi + h ,\tag{4.9}$$

where a is the bifurcation parameter and g is the parameter of the diffusion term.

As noted above, the medium under study is initially divided into domains where the order parameter has opposite signs. A new equilibrium in the system under the influence of the field displaces the borders between domains. As a result, the domains can disappear or at least decrease. As is well known, the spherical domain radius R changes in time:

$$\frac{\mathrm{d}R}{\mathrm{d}t} = -\frac{2g}{R} + \frac{\sqrt{bg}}{a}h .\tag{4.10}$$

As the field h grows, decreasing domain sizes correspond to increasing correlation radius r_c. This means that, averaging over the whole investigated bulk, the correlation radius will change, i.e., it oscillates by an amount δr, where $r_c = r_{c0} + \delta r$.

Taking into account (4.8) as well as the fact that the adiabatic compressibility $\beta_S = K(r_c/\sigma)$, where $K = \text{Const.}$, one can understand the influence of the ultrasound field on the amplitude of critical absorption.

To some extent this result elucidates the role of an external perturbation in systems close to the instability region. In Chaps. 7 and 8, we will consider in detail the influence of external fields on a binary liquid system close to the phase-separation region in a non-equilibrium state. It is important to find out when an external perturbation favors the ordering process and when it impedes it.

Because of the deviation from T_c, the system may lie within a narrow region of metastable states, where a liquid can be considered as a system in which a spherical nucleus of radius R_c of one of the components of a binary mixture has formed locally. The critical radius for such a nucleus to be in equilibrium with the surrounding liquid is determined by the well known relationship (see [106], for example)

$$R_c = \frac{2\sigma}{\Delta\mu} ,\tag{4.11}$$

where $\Delta\mu$ is the difference between the chemical potentials of the main solution and the substance of the nucleus. The difference [99]

$$\sigma \sim \sigma_1 - \sigma_2 \tag{4.12}$$

is known to great accuracy for many solutions, where σ_1 and σ_2 are the surface tensions of the pure components. Comparing (4.11) and (4.12) shows that the bigger the difference between the surface tensions of the pure solution components, the larger is the nuclear radius. The appearance of a nucleus in a homogeneous system gives rise to structural degradation and changes viscous effects in the liquid. Thus, as the nuclear radius grows, the viscosity coefficients increase and this in turn conditions the growth of the absorption amplitude (4.8).

Actually, experimental results show that there is a correlation between sound absorption and the difference of surface tensions of the pure components, and the greater this difference, the higher the absorption [107].

4.4 Chemical Reactions in Near-Critical States

Experimental data on chemical reactions occurring at a critical phase-separation point (CPS) are contradictory [108]. Let us consider the main statements and conclusions of existing theories.

In [108] the authors studied a binary liquid in which the reaction $B_2 \rightleftharpoons 2B$ was proceeding. It is known [109] that, at small deviations from equilibrium, the rate of this reaction is proportional to the chemical affinity A described by the equation

$$A = -\sum v_i \mu_i , \tag{4.13}$$

where μ_i is the chemical potential and v_i is the stoichiometrical coefficient of the reaction. For the considered reaction, $A = 2\mu_B - \mu_B$. Supposing the degree of reaction completeness χ to be determined by the equation $v_i d\chi = dN_i$ (where dN_i is the change in the number of moles of the ith component due to the reaction), for a small deviation of χ from equilibrium, we have

$$\frac{d\chi}{dt} = zA , \tag{4.14}$$

where $d\chi/dt$ is the reaction rate and z is a constant. By expanding A in powers of χ for constant thermodynamical parameters, the following equation was obtained [108]:

$$\frac{d\chi}{dt} = z \frac{dA}{d\chi_{eq}} (\chi - \chi_{eq}) . \tag{4.15}$$

It was also established that, regardless of the chemical nature of the components, the derivative $dA/d\chi$ changes as follows while approaching the critical point:

$$\frac{dA}{d\chi} \propto [(T - T_c)/T_c]^\gamma , \tag{4.16}$$

where γ is the critical index. When all components composing a solution take part in the reaction, $\gamma = 1.25$. When the solution consists of only one component, $\gamma = 0.12$.

The report [110] demonstrates the inaccuracy of the conclusions reached in [108]. The experimental results of I.P. Krichevsky, taken as the basis for the theory in [108], were satisfactorily explained by diffusion of light near a critical point [110]. Having pointed out that the divergence of (4.16) must be much weaker than predicted by the theory [108], J.C. Wheeler did not quantitatively evaluate γ because of the insufficiency of the available experimental material.

4.5 Kinetics of Mono- and Bimolecular Reactions Close to a Phase-Separation Critical Point

Using acoustic experiments, we made an attempt to reveal changes in the rates of chemical reactions close to a liquid–liquid phase-separation critical point. Two systems were selected as research subjects. One of these was water–cyclohexanol–methanol, where the observed relaxation is connected with the conformation transition in the cyclohexanol molecule. The non-critical relaxation process in this system is not driven by diffusion. A diffusion-driven reaction proceeds in a heptane–acetic acid–water solution, which is conditioned by the dimerization process, and moderation of the diffusion at the critical point may essentially affect the rate of the chemical reaction. To obtain information on chemical contributions to sound absorption from the data on total sound absorption, we suggested the following: contributions to the total sound absorption connected with concentration fluctuations and with non-critical processes are additive values, so that the FB scaling function [the absorption function $\alpha_p/\alpha_{max} \sim I(\omega)$] determines the frequency and temperature dependence of the sound absorption accurately enough.

Critical absorption was determined in accordance with the relaxing heat capacity theory by (4.5). The non-critical part of the absorption connected with rotation isomers or dimerization is determined by [111]

$$\frac{\alpha_p}{\omega^2} = \frac{A}{1 + (\omega/\omega_r)^2} + B, \qquad (4.17)$$

where A is the relaxation amplitude, B is the contribution to absorption of other factors independent of this relaxation process, and ω_r is the relaxation frequency of the characteristic process, all three being experimentally determined values.

Note that, in order to process the experimental acoustic data within the framework of the relaxation heat capacity theory, the shear viscosity and density of solutions were first measured in the necessary temperature range.

It was established by spectroscopy methods [112] that cyclohexanol molecules may have two advantageous conformations of the 'chair' form, with equatorial hydroxyl and axial groups. The broken balance between these conformations during ultrasound propagation leads to additional absorption.

The absorption coefficient and the sound velocity in pure cyclohexanol have been studied rather thoroughly [113]. Two regions of acoustic relaxation, low frequency (104–106 Hz) and high frequency (108–1 010 Hz), are observed

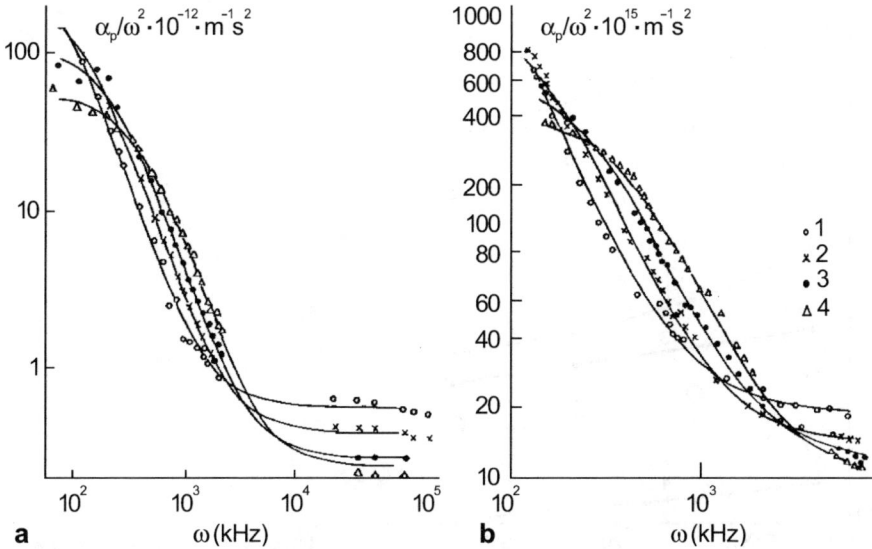

Fig. 4.3. Frequency dependence of sound absorption in cyclohexanol (**a**) and 1.4% cyclohexanol–water solution (**b**) at temperatures (1) 303 K, (2) 313 K, (3) 323 K, (4) 333 K

in the frequency range 20 kHz–3 GHz and in the temperature range 298.15–328.15 K.

To study the effects of a solvent on the kinetics of conformation transitions, we investigated a 1.4% cyclohexanol–water solution and pure cyclohexanol. Measurements of α_p/ω^2 in the frequency range 0.1–150 MHz at temperatures 303.15–333.15 K are shown in Figs. 4.3a and b. The processed experimental curves display an insignificant lowering of relaxation times (30%) in strongly diluted cyclohexanol compared with a pure substance. This attests to the weak impact of intermolecular interactions on the relaxation process. Figure 4.4 shows the temperature dependence of the non-critical relaxation process parameters: A (dashed lines) and B (solid lines). They are well approximated in the investigated temperature range by [35]

$$A = A_0 \exp \frac{a}{T}, \quad \omega_p = \omega_0 \exp \frac{d}{T}, \quad B = B_0 \exp \frac{b}{T}, \tag{4.18}$$

where a, b, d and A_0, ω_0, B_0 are constants.

The critical point in the water–cyclohexanol–methanol mixture was approached by changing the temperature. The experiment involved a large number of temperature cross-sections, a wide frequency range, and a high experimental accuracy to separate critical and non-critical contributions to the sound absorption using the least squares method and to determine the relaxation frequency ω_r and constants A and B at various temperatures. These constants (black points in Fig. 4.4) are characterized by the same tem-

48   4. Dynamics of States Close to Critical

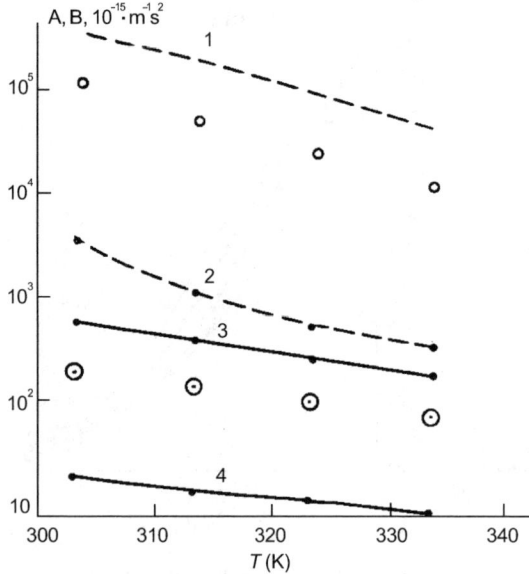

Fig. 4.4. Temperature dependence of A and B

perature dependence for pure cyclohexanol (A curve 1, B curve 2) and for a non-critical concentration of 1.4% cyclohexanol–water solution (A curve 3, B curve 4), and have intermediate values (spotted circles). The expected relaxation parameters calculated by (4.17) coincide with those obtained experimentally within the range of possible errors (Table 4.1). This means that critical phenomena do not significantly influence the relaxation process connected with conformation transitions in cyclohexanol.

The analysis also demonstrates that analogous research may be carried out for more complicated systems, where some influence of critical fluctuations on the relaxation process is more probable. Acetic acid and its solutions are subjects where acoustically observed relaxation in the range 0.05–70 MHz is conditioned by the establishment of equilibrium in the dimerization reaction [35, 111].

Table 4.1. Relaxation parameters A, B and ω_r for monomolecular reactions in cyclohexanol

Substance	A $[10^{-15}\ \mathrm{s}^2/\mathrm{m}]$	B $[10^{-15}\ \mathrm{s}^2/\mathrm{m}]$	ω_r [kHz]	T [K]
Cyclohexanol	209 700	422	99	313.14
1.4% cyclohexanol–water	1 222	15	129	313.14
Solution with critical concentration x_c	35 581	125	110	319.5

In the mixture heptane–acetic acid–water, the critical points were approached by small changes in the concentration of water. Experiments carried out far from a critical point proved that small additions of water do not significantly influence the acoustic relaxation parameters connected with the dimerization of acetic acid molecules.

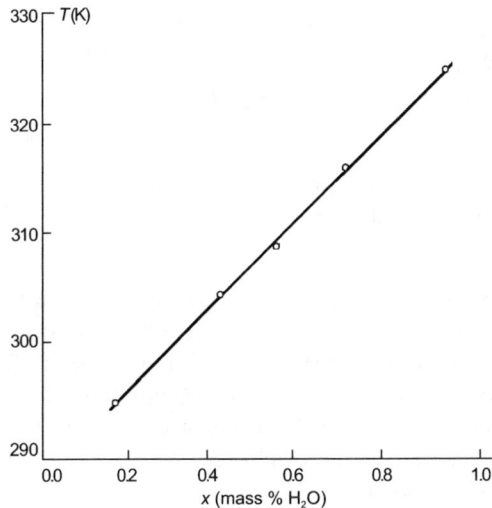

Fig. 4.5. Dependence of the critical separation temperature on the amount of water in solution

The dependence of the critical separation temperature on the amount of water is close to linear (Fig. 4.5). Therefore, we did not use the reduced temperature t in calculations with (3.5), but the related variable

$$x = z\left(\frac{x}{x_c} - 1\right),$$

where x_c is the critical water concentration and z is a constant calculated from the phase diagram and characterizing the degree of approach towards a critical point. The frequency dependence of the sound absorption for solutions of acetic acid in heptane (curve 2) and for the same solution with 0.0108 m.f. water (curve 1) are shown in Fig. 4.6. Processing the experimental data in the vicinity of the critical point by means of (3.5) and (4.17), we were able to separate critical and non-critical contributions to the sound absorption.

Within the range of experimental error, the parameters of the non-critical relaxation process turned out to have the same values as those for a waterless heptane–acetic acid solution. Table 4.2 displays the acoustic parameters of the acetic acid–heptane solution and those of the same solution with addition of water, found using the various equations.

Experimental data for the solution containing intermediate water concentrations (not displayed in Table 4.2) were processed by (4.5) and (4.17). The solid line numbered 3 in Fig. 4.6 (non-critical part of absorption α_p/ω^2)

4. Dynamics of States Close to Critical

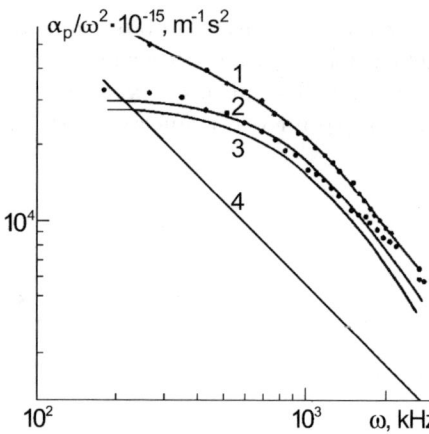

Fig. 4.6. Frequency dependence of sound absorption

and solid line 4 (critical part of absorption $\alpha_{\max}/\omega^2 - B$) are the results obtained using (4.6) and (4.17) (proceeding from the parametric constancy of the non-critical relaxation process) for the solution containing 0.0108 m.f. water. The experimental data are in a good agreement with calculated values and this attests to the fact that the non-critical relaxation process, as in the previous system, does not depend on the degree of closeness to the critical point, within the range of experimental error.

Table 4.2. Relaxation parameters A, B (10^{-15} s^2/m) and ω_r (kHz) for a bimolecular reaction in acetic acid obtained by (4.5) and (4.17)

Water	A		B		ω_r	
[m.f.]	I	II	I	II	I	II
0	34 600	24 400	300	250	954	1 104
0.0108	28 350	24 400	250	250	1 091	1 104

The rate of a chemical reaction is known to depend on the speed at which particles of reagents draw together. In a gaseous phase, this rate is determined by the number of collisions per unit time, and hence by the velocities of the particles. In a solution, the movement of particles with respect to each other is diffusive and a solvent presents an obstacle to their free movement. Therefore, the frequency with which reactionable particles collide with each other is limited by the diffusion rate of reagents.

If two particles necessarily react with each other at every collision, the rate of the respective chemical processes is determined only by the rate at which these particles are able to diffuse towards each other. If in a solution the diffusion goes slowly or tends to zero, as is observed near DCP [115, 116], it conditions the reaction rate constant, since the resulting rate of the overall process is always limited by the slowest of the contributing processes. Such reactions are called diffusion-driven processes. As shown in [117], the diffusion equation implies the following relation for the rate constant k:

$$k = 4\pi D_{AB} R_{AB} , \qquad (4.19)$$

where $D_{AB} = D_A + D_B$, $R_{AB} = R_A + R_B$, with D_A, D_B the diffusion coefficients of reagents A and B, and R_A, R_B the molecular radii.

Let us consider the approximation (4.19) regarding the objects under investigation. Data obtained for carbonic acid, one of the first liquids [118] where an ultrasound relaxation process was observed, are very complicated and contradictory. In the solution, a relaxation process occurs, connected with establishment of equilibrium in the dimerization reaction. For acetic acid, this reaction has the form [35]:

$$2\,CH_3C\overset{\displaystyle O}{\underset{\displaystyle OH}{}} \;\rightleftharpoons\; \overset{\displaystyle O...HO}{\underset{\displaystyle OH...O}{CCH_3 \quad CCH_3}}$$

where the forward reaction is of second order and the reverse reaction is of first order.

In order to evaluate the reaction rate constants quantitatively, one needs data on vapor pressure and IR spectroscopy. These were obtained in [119] for a solution of acetic acid in carbon tetrachloride. For temperature 293 K, the results of our processing are as follows:

- the forward reaction rate constant $k = 2.16 \times 10^{-8}$ molecules/s,
- the opposite reaction rate constant $k = 0.492 \times 10^{-5}$ molecules/s,
- the enthalpy of activation of the forward reaction is 10 032 J/mol,
- the enthalpy of activation of the reverse reaction is 48 738 J/mol.

As can be seen, k lies in the range 10^{-8}–10^{-5} molecules/s and, according to (4.19), the dimerization reaction in acetic acid must depend on the diffusion rate. Therefore, the kinetic parameters of the bimolecular reaction depend on the proximity of the critical point.

Note that the theory of diffusion-limited reactions [120] describes chaotic displacements of the dissolved substance molecules well when they are far from each other. However, if two molecules of the dissolved substance are separated by one or two molecules of the solvent, their displacements may become interdependent. It is of great importance to take this fact into account when considering solutions close to a critical point of separation. In these systems, fluctuation effects prevail near this point, and at distances shorter than the correlation radius, the system becomes homogeneous, resulting in the slowing of diffusion processes driven by the difference in concentrations at different points of the space. This should lead to lowering of the reaction rate constant k. However, as experimental results show, it actually remains constant. This seems strange to say the least.

The point is that a change in the reaction rate constant is determined by a change in the diffusion coefficient that follows from (4.19). The seeming contradiction can be explained if we remember that k near CPS depends not only on the radii R_{AB} of interacting molecules, but also on the correlation radius r_c. As mentioned above, the value of r_c grows when approaching a critical point, and if the rate of change in correlation radius $\mathrm{d}r_c/\mathrm{d}t$ compensates changes in the rate of the diffusion coefficient $\mathrm{d}D/\mathrm{d}t$, the chemical reaction rate will be constant.

Moreover, as shown by several evaluations [14], one can neglect the diffusion contribution to the critical sound absorption up to $t \approx 10^{-4}$.

Investigations have thus shown that relaxation processes connected with both an intramolecular conformation transition and a dimerization reaction occur independently of the degree of approach to a phase transition.

5. Physics of Solutions with Double Critical Point

The state of a system in the region where upper (UCPSP) and lower (LCPSP) critical phase-separation points merge has been named a double critical point (DCP). This term is not universally accepted. Other terms such as a hypercritical or reentrant point are also employed in the literature.

Studies of the systems whose UCPSP and LCPSP coincide are of great interest for many different reasons:

- the existence of such systems, in which these two critical points (CP) merge into one (DCP) under a change in pressure, admixture content or other system state parameters, is an unusual and remarkable fact,
- by investigating the DCP, we may discover the character and interaction mechanisms of second-order phase transitions,
- such systems are unusually susceptible to various external forces that create favorable conditions for the most complete and exact check of fluctuation theories [14, 46],
- some theories have recently appeared [45, 59, 63] that can reliably interpret a large number of experimental data within the framework of the phenomenological approach for any systems characterized by the existence of two closely located points of the second-order phase transition.

Despite the fact that a double critical point lies on the line of critical points, the state of a system in DCP has new specific peculiarities. Effective critical determining properties of systems close to phase transition depend on both the solution temperature and the width of the closed region between the UCPSP and LCPSP, with critical indices doubling near DCP. Some properties of systems with a DCP (e.g., osmotic compressibility, light scattering, correlation radius) have anomalously high divergence, others (e.g., heat capacity) do not have any anomalies at the DCP. Another peculiarity of such systems is that they allow one to study the critical behavior of solutions with a DCP depending not only on temperature but also on the concentration of admixtures. Experimental investigations on solutions with a DCP extend our understanding of the nature of critical phenomena, establish a basis for studying phase transitions of higher order in multicomponent solutions, and reveal perspectives for practical applications of the obtained results, in particular, when seeking to design sensors for external parameters.

5.1 Theory of Solutions with Double Critical Point

The scale invariance theory of systems with a double critical point, based on the principle of metamorphism of critical phenomena, is elaborated in [14]. It is established that the behavior of thermodynamic values for the points on the CP line is determined from the isomorphous thermodynamic potential. Calculations show that if, as a result of a change in temperature, a system approaches a given critical point with temperature $T_c(h)$, the isomorphous thermodynamic potential can determine the behavior of thermodynamic values for the points on the CP line direction close to the normal of the CP line. The power dependence of various physical values on the difference $T - T_c(h)$ is characterized by the constancy of values of critical indices if the system approaches the point $T_c(h)$ along the tangent to the CP line. Critical exponents of the order parameter β, compressibility γ, correlation radius ν, and so on, all double. Systems with double critical point are among those whose temperature approaches $T_c(h)$ along the tangent.

The theory of systems with a DCP was further developed in [63], where the square-root dependence of $T_c(h)$ at the DCP was revealed to reflect the parabolic temperature dependence of the coefficient α in the Landau free energy functional:

$$F = F_0 + \int \left[\frac{\alpha(T,h)}{2} \varphi^2 + \frac{\beta(T,h)}{4} \varphi^4 + \ldots + \frac{g}{2}(\nabla\varphi)^2 \right] d\nu . \quad (5.1)$$

The coefficient α at each critical point proves to be zero so there must be a minimum between these points:

$$\alpha(T,h) = \alpha_m(h) + \alpha_2(h)[T - T_m(h)]^2 . \quad (5.2)$$

Provided that $|T - T_m|/T \ll 1$ and the highest expansion terms $|T - T_m|/T$ can be neglected, the temperature $T_m(h)$ is determined by

$$T_m(h) = \frac{T_B(h) - T_H(h)}{2} , \quad (5.3)$$

where T_B and T_H are the temperatures of the UCPSP and LCPSP, respectively. The coefficient α can be written as

$$\alpha = \alpha_2(h)[T - T_B(h)][T_H(h) - T] . \quad (5.4)$$

Then from (5.1), (5.2), (5.4), it follows that

$$\varphi^2 = -\frac{\alpha}{2b} = \frac{\alpha_2(h)}{2b}[T - T_B(h)][T_H(h) - T] . \quad (5.5)$$

If $[T_B - T_H]/[T_B + T_H] \ll G$, where G is the Ginsburg number [14], the whole region of phase separation is localized inside the fluctuation region. In this case, the functional is modified as follows [63]:

$$F = F_0 + \int \left[\frac{a_0}{2} \left| \frac{\alpha}{\alpha_c} \right|^{2\nu-z-1} \varphi^2 + \frac{b_0}{4} \left| \frac{\alpha}{\alpha_c} \right|^{\nu-2} \varphi^4 + \frac{g_0}{4} \left| \frac{\alpha}{\alpha_c} \right|^{-z} (\nabla \varphi)^2 \right] d\nu ,$$

(5.6)

where $z \approx 10^{-2}$ is a small critical index characterizing the behavior of the correlation function at $\alpha = 0$ and

$$\alpha_c = \left(\frac{3k_B T b}{8\pi g^{3/2}} \right)^2 .$$

For the coexistence curve, we have in this case

$$\varphi \sim [(T_B - T)(T - T_H)]^\beta . \tag{5.7}$$

The correlation radius depends on temperature as follows:

$$r_c \sim [(T_B - T)(T - T_H)]^{-\nu} . \tag{5.8}$$

The parabolic form of the function $\alpha(T)$ in the expression for the thermodynamic potential has been successfully used [63] in the qualitative description of the anomalous temperature behavior of many Seignette salt properties.

5.2 Rayleigh Scattering of Light

It is known that, when a substance approaches DCP, one observes anomalous light scattering called opalescence. The main laws of Rayleigh light scattering, which are observed in ordinary states, are found to be broken in this case. The anomalous absorption of light is conditioned by an increase in fluctuations of the refractive index which in its turn is connected with density fluctuations, and concentration or inter-orientation of anisotropic molecules [121, 122].

The theory of critical opalescence was developed by J. Ornstein and P. Zernike [123]. In the Ornstein–Zernike approximation, in the vicinity of a critical point, we have for the light-scattering coefficient R_θ, at the scattering angle θ,

$$R_\theta^{-1} = R_0^{-1} + R_0^{-1} r_c^2 q^2 , \tag{5.9}$$

where R_0 is the light-scattering coefficient at zero scattering angle, q is a wave vector,

$$q = \frac{4\pi}{\lambda} \sin \frac{\theta}{2} , \tag{5.10}$$

and λ is the wavelength of the transmitted light. We have used the fact that the incident light is vertically polarized, and the scattered and incident light lie in a horizontal plane. From (5.10) one can draw the following conclusion: the dependence of the reciprocal value of the light-scattering coefficient R_θ^{-1} on $\sin^2(\theta/2)$ is described by a straight line cutting off a section on the abscissa equal to the reciprocal value of the light-scattering coefficient at zero angle. In the Ornstein–Zernike approach, the magnitude $R_0^{-1} r_c^2$ proves to be constant.

Numerous experimental investigations [124–126] confirm the conclusion that R_θ^{-1} depends linearly on $\sin^2(\theta/2)$. According to the theory of scale invariance,

$$R_0 \sim \beta_{oc} \sim |T - T_c|^\gamma , \qquad (5.11)$$

where β_{oc} is the osmotic compressibility of the solution, and γ is the critical index, equal to 1.23 ± 0.01.

When approaching a critical point, the correlation radius increases according to

$$r_c = r_0[(T - T_c)/T_c]^\nu , \qquad (5.12)$$

where r_0 is the short-range correlation radius.

Thus, in the Ornstein–Zernike approach the indices γ and ν are coupled by $\gamma = 2\nu$. In a more exact description of critical phenomena, a critical exponent ξ is introduced which is the exponent of anomalous behavior of the correlation function distinguishing it from the Ornstein–Zernike approach. The relation between exponents γ and ν then takes the form

$$\gamma = \nu(2 - \xi) . \qquad (5.13)$$

The magnitude of ξ is very small, and the relation (5.13) is transformed to the formula $\gamma = 2\nu$, with accuracy up to 1–2%. For further description of critical light scattering, we use the Ornstein–Zernike approach, because the accuracy of experimental determination of critical indices is comparable with the value of ξ.

In this approach

$$R_0 = A_1[(T - T_c)/T_c]^\gamma , \qquad (5.14)$$

where A_1 is a fixed coefficient. Thus, by employing the Ornstein–Zernike approach, it is possible to determine both critical indices γ and ν and the short-range correlation radius r_0 from the angular and temperature dependence of the light-scattering coefficient.

One of the first works on the Rayleigh scattering of light in a solution with a double critical point was [125], where a solution of β-picoline–water + heavy water was the subject of investigation. Measurements showed that isotherms of the light-scattering coefficient of the binary system β-picoline–water and the ternary system β-picoline–water + heavy water are very similar. The maximal intensity of the light scattering is located nearly in the same range of concentrations of β-picoline, and the intensity of light scattering in the ternary system is much higher than in the binary system. This is due to the growth of concentration fluctuations when substituting some fraction of water by heavy water. Such an increase in the intensity of scattering is natural, because the system begins to separate at a certain concentration of heavy water.

5.2 Rayleigh Scattering of Light

Scattering of light was studied most completely while investigating solutions of glycerine–guaiacol + water [62, 126, 127]. In this system, water is not one of the main components, but an additive on account of which a closed separation region is formed. The reports [62, 127] present investigations of the line width of Rayleigh light scattering as a function of the width $\Delta T = T_B - T_H$ of the separation region. When processed, the experimental data showed [62, 127] that the obtained values of the critical exponents are close to double. However, a complete doubling was not observed.

The authors of [59] investigated the angular dependence of the intensity of scattered light as a function of the width of the separation region, including the absolute homogeneity region. The effective index $\gamma' = 2\gamma = 2.48$ was observed to double near DCP. Thus, conclusions reached in [62,127] and [126] are, in principle, contradictory.

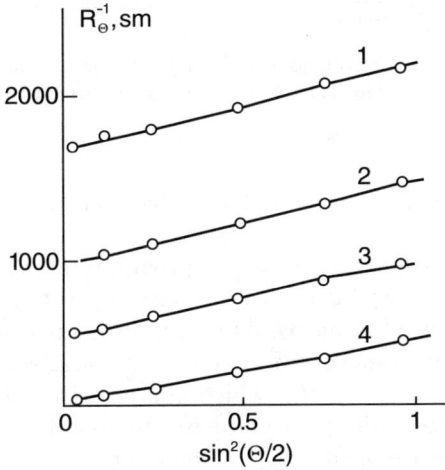

Fig. 5.1. Angular dependence of light scattering for various values of ΔT: (curve 1) 10 K, (curve 2) 8.1 K, (curve 3) 5.9 K, (curve 4) 0.6 K. Temperature dependence of A and B

To study the dependence of light scattering on the scattering angle, temperature, admixture concentration and pressure, solutions of butanol–water + HCl, propanol–water + NaCl, and methylethylene–water + 1,4-dioxane were investigated in [59]. The angular dependence of light scattering at various values of ΔT, in the coordinates R_θ^{-1} and $\sin^2(\theta/2)$, proved to be characterized by straight lines of the same slope (Fig. 5.1). This completely confirms the applicability of the Ornstein–Zernike formula for solutions close to DCP.

In order to elucidate the applicability of (5.14), studies were carried out to determine the dependence of the light-scattering coefficient on proximity t of the solution temperature to the critical point temperature.

Values of R_0 were found by extrapolation of experimental data to zero scattering angle. The experimental dependence of $\ln(R_0/A_1)$ on $\ln t$ is shown in Fig. 5.2. It should be noted that this dependence deviates from a straight line, the deviation significantly exceeding experimental accuracy. This at-

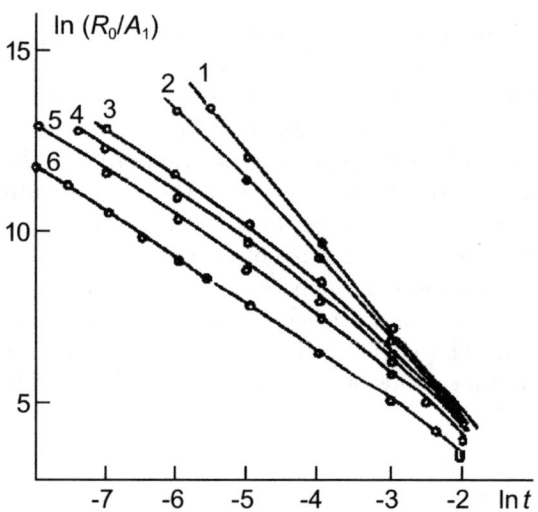

Fig. 5.2. Dependence of $\ln(R_0/A_1)$ on $\ln t$ for solutions with a DCP for various ΔT: (curve 1) 0, (curve 2) 0.029, (curve 3) 0.035, (curve 4) 0.04, (curve 5) 0.078, (curve 6) 0.196

tests to the fact that the critical exponent γ becomes a variable, depending significantly on both t and ΔT.

It is natural to assume that the above is caused by the proximity of two phase transitions with upper and lower critical points of phase separation, which hold the system under their mutual influence. One can also suppose that critical properties near DCP are functions of two independent variables, namely $t_1 = (T_B - T)/T_0$ and $t_2 = (T - T_H)/T_0$, which characterize the closeness of the system to a critical state with upper and lower temperatures of phase-separation (T_0 is the temperature of the DCP). Therefore,

$$R_0 = K t_1^\gamma t_2^\gamma . \tag{5.15}$$

Analysis shows the universality of (5.15), describing the temperature dependence of the light-scattering coefficient in the whole vicinity of DCP with a high level of accuracy (Fig. 5.3). Values of $\ln(R_0/A_1)$ are seen to lie on the same straight line with slope $\gamma = 1.24 \pm 0.03$ at various values of $\ln(t_1 t_2)$.

Thus, the universality of the relation (5.15) confirms the assumption of superposition of critical phenomena in the region of DCP due to phase transitions of solutions with upper and lower critical points of phase-separation. In addition, (5.14) and (5.15) are completely analogous, provided that

$$t = t_1 t_2 . \tag{5.16}$$

The relation (5.15) can be generalized to include the effect either of changes in temperature or any other external parameter H on the behavior of a system. We write the expression (5.15) as

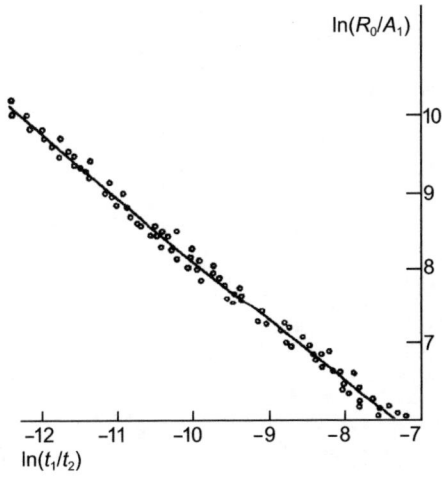

Fig. 5.3. Dependence of $\ln(R_0/A_1)$ on $\ln(t_1 t_2)$ for solutions with a DCP

$$R_0 = K(t_0^2 - \Delta t_0^2)^{-\gamma}, \tag{5.17}$$

where $t_0 = (T - T_0)/T_0$, $\Delta t_0 = (T_B - T_H)/T_0$. Further, by taking into account the expression

$$M\left[\frac{T - T_0}{T_0}\right]^2 = \frac{H - H_0}{H_0}, \tag{5.18}$$

where H_0 is a generalized parameter of the force corresponding to DCP and M is a constant coefficient, we obtain

$$R_0 = K\left[\frac{H - H_0}{MH_0} - \Delta t_0^2\right]^{-\gamma}. \tag{5.19}$$

In the particular case $\Delta t_0 = t_1$,

$$R_0 = K\left[\frac{(H - H_0)}{MH_0} - \frac{(H_1 - H_0)}{MH_0}\right]^{-\gamma} = K\left[\frac{(H - H_1)}{MH_0}\right]^{-\gamma}, \tag{5.20}$$

where H_1 is the external force, necessary for separation of the solution at the temperature $\Delta t_0 = t_1$.

It turns out that (5.17) and (5.20) describe the dependence of the light-scattering coefficient not only in the vicinity of the upper, lower and double critical points, but also in the region of absolute solubility, i.e., of a so-called specific point.

Figure 5.4 shows the results of processing the experimental data with (5.19). Curves 4–7 with a finite maximum correspond to the region of absolute solubility, the divergent curve 3 corresponds to DCP and curves 1 and 2 correspond to the phase-separation region. Curves showing the dependence of the light-scattering coefficient on pressure P at various temperatures are given in Fig. 5.5 [128]. After processing the experimental data (solid lines) with (5.20), the critical index was found to be $\gamma = 1.24 \pm 0.02$.

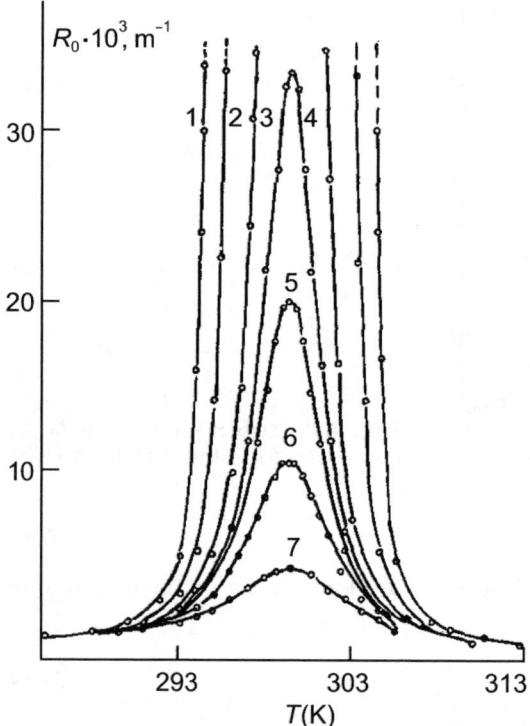

Fig. 5.4. Temperature dependence of the light-scattering coefficient

It should be noted that the temperature dependence of the intensity of scattered light in the region of absolute solubility can also be described by (5.17). In this case the last term of the equation must satisfy the condition $\Delta t^2 < 0$ which requires the introduction of upper and lower critical temperatures.

The appearance of imaginary upper and lower temperatures may attest to the presence of an imaginary phase diagram as a continuation of the real one into the region of absolute solubility, in which the lines of upper critical temperatures lie under those of the lower critical temperatures. This may imply the absolute solubility of solution components.

In the Ornstein–Zernike approximation, the short-range correlation radii of solutions close to DCP were found in various separation regions and in the state of absolute homogeneity. These radii turned out to be practically constant, independent of pressure, temperature and amount of admixture. Table 5.1 presents calculated values of the short-range correlation radius r_0 as well as values of the scaling coefficients. The constant magnitudes of the short-range correlation index and radius r_0 in a wide region of DCP, and

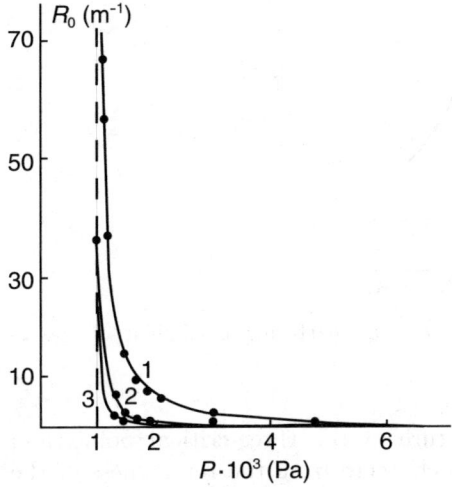

Fig. 5.5. Dependence of the light-scattering coefficient R_0 on pressure for solutions of (1) butanol–water + HCl, (2) propanol–water + NaCl, and (3) methylethylketone–water + 1,4-dioxane

their correspondence to the three-dimensional Ising model, confirm the isomorphous state of the thermodynamic variable $t = t_1 t_2$.

It is interesting to extend the superposition principle to three phase transitions present in the system 2-butanol–water + propanol under investigation. In accordance with (5.20), the coefficient of light scattering for three phase transitions can be expressed as

$$R_0 = K \left[\frac{H_{01} - H}{M H_{01}} - \Delta t_{01}^2 \right]^{-\gamma} \left[\frac{T - T_{c3}}{T_{c3}} \right]^{-\gamma} . \quad (5.21)$$

For comparison of theoretical data with experimental data obtained for the system of the given composition $x = x_{01} \neq x_{c3}$, where x_{c3} is the critical composition of the third and lowest temperature phase transition, we used the expression obtained from (5.21) by taking into account the simplified 'linear model' of the scale equation of a state [31]:

$$R_0 = K \left[\frac{H_{01} - H}{M H_{01}} - \Delta t_{01}^2 \right]^{-\gamma} \left[\frac{T - T_{03}}{T_{03}} \right]^{-\gamma} f_1(x/x_{c3}) , \quad (5.22)$$

$$r_c = r_0 (t_1 t_2)^{-\nu} f_2(x/x_{c3}) t_{03}^{-\nu} , \quad (5.23)$$

Table 5.1. Calculated values of the critical opalescence parameters for solutions of (A) 1-butanol–water + HCl, (B) 2-propanol–water + NaCl, and (C) 3-methylethylketone–water + 1,4-dioxane

Solution	$A_1 \times 10^8$ [cm^{-1}]	$r_0 \times 10^{10}$ [m]	γ	ν
A	14	4.1	1.23 ± 0.03	0.63 ± 0.02
B	3.7	3.1	1.23 ± 0.03	0.62 ± 0.02
C	1.8	2.9	1.23 ± 0.03	0.62 ± 0.02

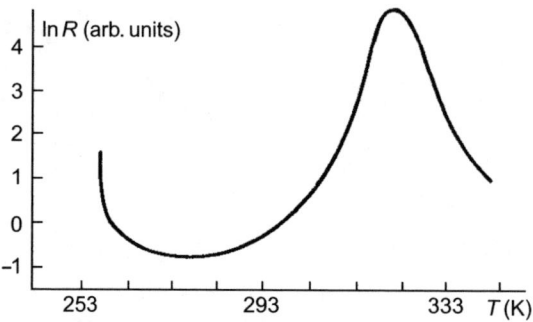

Fig. 5.6. Temperature dependence of the light-scattering coefficient of the 2-butanol–water + propanol solution

where $\gamma = 1.23$, T_{03} is the temperature of the three-critical point, and $f_1(x/x_{c3})$ and $f_2(x/x_{c3})$ are functions characterizing the remoteness of the system from the critical concentration of the components. Figure 5.6 shows the temperature dependence of the light-scattering coefficient of the 2-butanol–water + propanol solution.

5.3 Dynamics of Near-Critical States of Solutions with a DCP

When generalized to kinetic phenomena in the vicinity of critical points, the scale invariance hypothesis is called dynamical scaling [14]. This theory assumes that, besides a characteristic correlation radius r_c, the liquid has a characteristic relaxation time τ for concentration fluctuations. The relation between these two can be determined by assuming [14] that, in the vicinity of a critical point, heat is transferred by spherical 'droplets' of radius r_c. For the heat conductivity due to such droplets, one can write an expression analogous to that for diffusion of Brownian particles in a viscous medium:

$$D = \frac{k_B T}{6\pi \eta_s^2 r_c}, \tag{5.24}$$

where k_B is the Boltzmann constant. For the critical point, at the scale of order r_c,

$$\tau = \frac{r_c^2}{D}. \tag{5.25}$$

From (5.24) and (5.25), we obtain

$$\tau = \frac{6\pi \eta_s^2 r_c^3}{k_B T}. \tag{5.26}$$

It should be noted that, in dynamical scaling, the shear viscosity follows the power law [14, 129]

$$\eta_s = \eta_p |t|^{-\phi} , \qquad (5.27)$$

where ϕ is the critical index of viscosity, and η_p is the regular part of the shear viscosity.

The reliability of (5.27) has been confirmed experimentally in [129, 130]. The critical index obtained is 0.04, in agreement with theoretical predictions from fluctuation theory. The strongest anomalies emerge when investigating the behavior of acoustic parameters near critical separation points. Theories describing the anomalous behavior of ultrasonic velocity and absorption in the critical state are presented in Chap. 3.

5.4 Shear Viscosity

Investigations into the temperature and concentration dependence of the shear viscosity η_s make it possible to justify the predictions of dynamical scaling in the vicinity of DCP, and to calculate magnitudes of characteristic relaxation times of concentration fluctuations (via data on the correlation radius). Universality of the superposition principle of phase transitions in the vicinity of the DCP has been confirmed in [59, 65].

The first experimental investigations into the shear viscosity of solutions near a DCP were carried out on 3-methylpyridine–water + heavy water [129], glycerine–guaiacol + water [62] and propanol–water + NaCl [33]. The values for the shear viscosity in the first solution were processed using (5.27), whilst for the second and third solutions, the theory of interacting modes was used [130, 131]. The critical index of viscosity was observed to range from 0.04 to twice the value depending on the magnitude of the separation gap ΔT. A change in the index contradicts the universality hypothesis for critical indices, viz., their constancy for the proper (isomorphous) choice of thermodynamic variables.

The viscosities of butanol–water + HCl, propanol–water + NaCl, and 2-butanol–water + propanol solutions were studied in [33, 41]. The viscosity was measured in terms of the temperature at $x = x_0$ for various values of ΔT in the solutions. In the viscosity curves (Fig. 5.7), one can see two pronounced maximums to the lower and upper separation points. As ΔT decreases, the two maximums combine into one, corresponding to a DCP.

It should be noted that the total width of the fluctuation region with two maximums increases when approaching the DCP (ΔT decreasing). Far from the separation points, the dependence of the logarithmic viscosity on the reciprocal temperature is described by a straight line, i.e., a temperature dependence of Arrhenius type is observed here.

The thermodynamic variables of the solutions under investigation are the temperature and also the concentrations of admixtures. The latter can be viewed as sources of the field changing the system structure and conditioning phase transitions. To elucidate the above, we investigated the behavior

64   5. Physics of Solutions with Double Critical Point

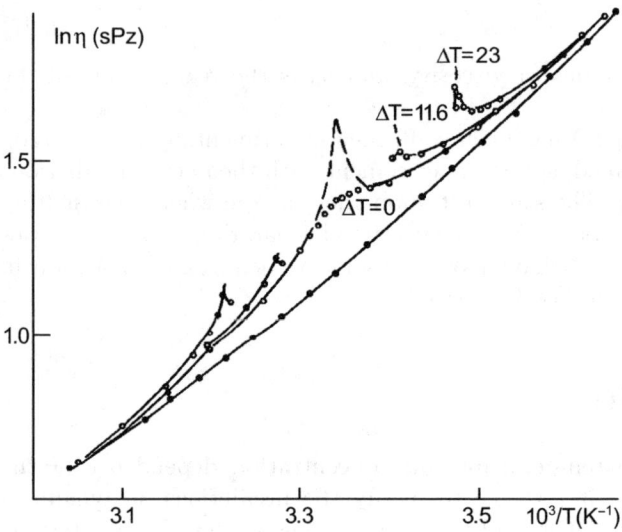

Fig. 5.7. Temperature dependence of shear viscosity for solutions with a DCP, for various values of ΔT: 0 K, 11.6 K, 23 K

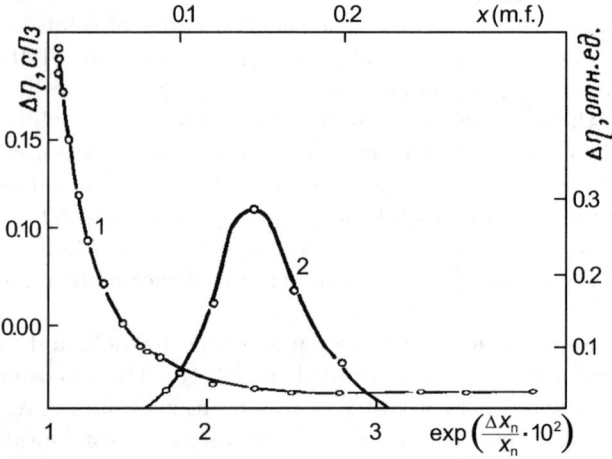

Fig. 5.8. Dependence of the excess part of the viscosity $\Delta \eta$ on the admixture concentrations in the vicinity of the DCP (1) and dependence of the maximal values of the excess viscosity on the propanol concentration in the solution (2)

of the viscosity as a function of the concentration of admixtures x_n. It was established that the viscosity has an anomalously sharp increase near x_{cp}, the critical concentration of the propanol. However, near the DCP, its experimental values remain finite. (The factors limiting growth of the viscosity near the DCP will be considered below.) The maximal, experimentally detected value of the excess viscosity $\Delta \eta$ equals 15%, compared with its regular part η_p. When the solution composition is altered by a change in the concentration x, the curve $\Delta\eta$ passes through a narrow maximum at the propanol concentration x_{cp} (Fig. 5.8).

These results were firstly analyzed within the framework of the theory of interacting modes [100, 132], which predicts divergence of the viscosity near a critical point:

$$\frac{\eta_s - \eta_p}{\eta_p} = \ln B - \phi \ln t, \tag{5.28}$$

where $B = q_0 r_0$ is a constant and q_0 is a constant with the dimensions of reciprocal length.

The viscosity measurements were analyzed using (5.28) for a range of values of η in which the secondary effect responsible for the finiteness of the viscosity values at the CP, related to the interaction of the concentration fluctuations and the emergence of the viscous flow velocity gradient upon the viscosity measurements [133,134], is not yet manifest. In our experiments this range proves to be $10^{-3} < t < 10^{-1}$ at $\Delta T \approx 0$ and at $\Delta T \to \alpha$ (the case of a single critical point). The effective index ϕ obtained for the solutions investigated was observed to increase as the width ΔT of the separation region decreased. The dependence $\phi(\Delta T)$ for propanol–water + NaCl (Fig. 5.9) turned out to be analogous to that obtained before [33] for guaiacol–glycerine + water with a DCP, viz.,

$$\lim \frac{\phi(0)}{\phi(\Delta T)} = 2. \tag{5.29}$$

Hence, one can talk about the doubling of the critical index at a DCP.

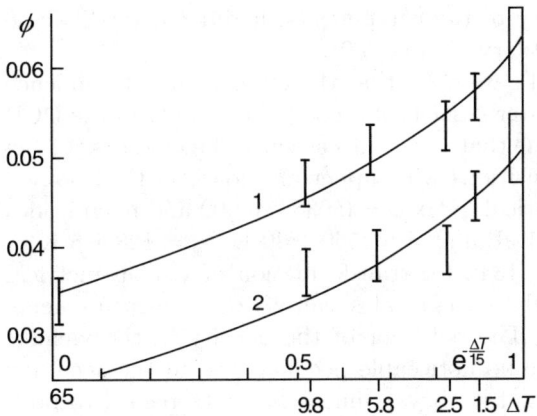

Fig. 5.9. Dependence of the effective viscosity index of LCPSP (1) and UCPSP (2) on the width ΔT of the separation region

The behavior of the excess viscosity $\Delta\eta/\eta$ of the system upon changing the admixture concentration $x_{\rm cp}$ is approximated by (5.22), in which $\phi(x_{\rm p}) = (x_{\rm p} - x_{\rm cp})/x_{\rm cp}$ is chosen as the thermodynamic variable.

The approximation range near the DCP corresponds to the values $10^{-3} < \phi(x_n) < 10^{-1}$, when $\eta = 0.034 \pm 0.004$ and $B = 0.9$. This shows that, when the system state is changed via some alterations in the concentration of admixtures, the experimental and theoretical values of the critical viscosity index overlap within the range of experimental error. Further, the viscosity measurements were analyzed using

$$\eta_{\rm s} = \eta_{\rm p} |t_1 t_2|^{-\phi}, \tag{5.30}$$

which follows from the theory of renormalization groups and which we have generalized for the case of superposition of two phase transitions in solutions with a DCP [41]. In (5.30), $\eta_{\rm p}$ is determined by the expression

$$\eta_{\rm p} = \eta_0 \exp \frac{W}{T + b_{\rm p}}, \tag{5.31}$$

where W is the activation energy of viscous flow and $b_{\rm p}$ is a coefficient characterizing the deviation of the dependence from one of Arrhenius type.

The results of approximation by (5.30) for butanol–water + NaCl (the solid line in Fig. 5.7) are in good agreement with experiment across a broad neighborhood of the DCP, except for the region in the immediate vicinity of critical temperatures, where the effect of phase-transition smearing is induced by the viscous flow necessary for shear viscosity measurements. The regular part of the viscosity, given by (5.31), was determined experimentally in a solution with a minor excess of HCl admixture compared with the concentration needed for obtaining DCP. As the viscosity of the solution was in the region of absolute solubility of the components, it did not in this case contain a singular component (curve 1, Fig. 5.9).

The processed experimental data show that the temperature dependence of the viscosity in a butanol–water + HCl solution in the vicinity of the DCP can be described by a power law that takes into account the closeness of the system state to two phase transitions with upper and lower critical points. The power law is set by the critical index $\phi = 0.038 \pm 0.002$ and magnitudes of the constants $\eta_0 = 2.35 \pm 0.1$ cP and $W = 560 \pm 40$ K, $b_{\rm p} = 128 \pm 5$ K.

It should be noted that in [134], for transformation of viscous-metrical measurements into values of the dynamical shear viscosity, measurements of density ρ were carried out. The behavior of the density in the vicinity of the DCP turned out to have no noticeable peculiarities (in the accuracy range 10^{-4} noted previously) and it obeys a linear law with regard to both temperature and concentration of admixtures.

5.5 Sound Propagation

One of the first objects for which the temperature dependence of ultrasonic absorption was studied at various concentrations ($x_n - x_{n0}$) of the admixture in the range of frequencies 4.5–135 MHz is butanol–water + Ca(CNS)$_2$ solution [52]. Here the following peculiarities appeared:

- the ultrasonic absorption diverges while approaching a critical point away from DCP,
- the anomaly of critical absorption significantly decreases in the vicinity of DCP and almost disappears there,
- the ultrasonic absorption abruptly increases at low temperatures, a fact which the authors assume to be caused by a phase transition of the first order, i.e., crystallization [52].

It should be noted, however, that the results presented in this work were not theoretically processed within the framework of the fluctuation theory and sound frequencies used in measurements were too high.

In [135], sound velocity and absorption were measured in a binary solution of 2-methylpyridine–heavy water via temperature and pressure at a frequency of 1 MHz. This solution has DCP at major component concentrations of 0.09–0.91 mol.frac., pressure 208 atm, and temperature 379 K. While measuring the absorption, the result was opposite to that found by [87]. A strongly pronounced maximum was revealed in the temperature region of the DCP, and the ultrasound velocity depended linearly on the temperature.

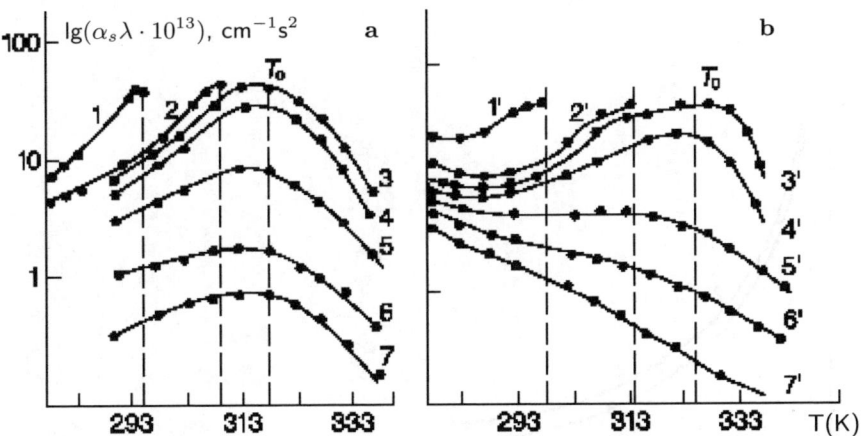

Fig. 5.10. Temperature dependence curves of the acoustic absorption of solutions propanol–water + NaCl (**a**) and 2-butanol–water + propanol (**b**) at various admixture concentrations: (curve 1) 5.877, (curve 2) 5.465, (curve 3) 5.383, (curve 4) 5.329, (curve 5) 5.003, (curve 6) 4.526, (curve 7) 3.819, (curve 1') 6.393, (curve 2') 8.901, (curve 3') 9.420, (curve 4') 10.0, (curve 5') 12.0, (curve 6') 14.0, (curve 7') 16.0

68   5. Physics of Solutions with Double Critical Point

To obtain detailed and unambiguous data on the behavior of α_s and ν_c in the vicinity of DCP, investigations [65] were carried out in the region of the lowest ultrasonic frequencies, i.e., at the frequency where critical absorption significantly increases. Propanol–water + NaCl and 2-butanol–water + propanol systems were chosen as the subjects of investigation, because the location of their DCPs is the most convenient to carry out experiments over a wide range of temperatures. In addition, the second system, as noted before, has a critical point of phase separation, which is located below a closed separation region.

Figure 5.10a gives the temperature dependence of acoustic absorption at the wavelength $\alpha_s\lambda$ of the propanol–water + NaCl solution at various admixture concentrations x_n in a wide region of DCP and at frequency 30 MHz. The temperature dependence curve has a maximum at the DCP, the value of $\alpha_s\lambda$ abruptly increases while approaching lines of critical points T, and the maximum on curves $\alpha_s\lambda(T)$ abruptly decreases and almost disappears while moving away to the region of values x_p corresponding to absolute solubility in the system.

It is characteristic that the value of critical ultrasonic absorption at given frequencies at the DCP and on the line of critical points $T(x_p)$ remains finite, reaching an amplitude of $\alpha_s\lambda$. All this attests to the fact that the observed anomaly of ultrasonic absorption in the vicinity of DCP is due to a relaxation process of the fluctuation type.

In fact, the frequency dependence of ultrasonic absorption near DCP detected a wide region of acoustic relaxation (Fig. 5.11). Acoustic measurements processed within the framework of FB dynamic scaling [101, 103, 136]

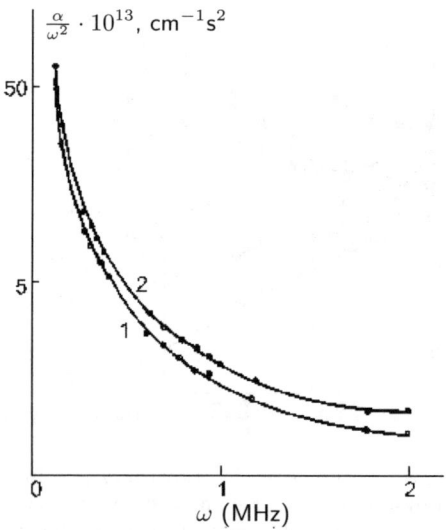

Fig. 5.11. Frequency dependence of ultrasonic absorption near DCP for solutions propanol–water + NaCl (1), 2-butanol–water + propanol (2)

(see Chap. 3) agree with theoretical predictions if the relaxation frequency of concentration fluctuations is a function of the proximity of the system temperature to the temperatures of two (propanol–water + NaCl) or three (2-butanol–water + propanol, Fig. 5.10b) phase transitions:

$$\omega_c \sim \prod_{i=1}^{n} t_i^{2-\alpha}, \tag{5.32}$$

where n is the number of phase transitions in the systems under investigation. In a homogeneous region, (5.32) takes the form

$$\omega \sim (t_1 t_2 t_3)^{2-\alpha}, \tag{5.33}$$

where $t_3 = (T - T_3)/T_3$ is the reduced temperature of the third phase transition (low temperature upper critical point). For the propanol–water + NaCl solution, $t_3 = 1$.

Experimental data were processed by the least squares method within the framework of the FB theory. Only one adjustable parameter, the short-range correlation radius, was used, making it possible to compare values obtained from optical and acoustic experiments. The solid lines in Fig. 5.10a are theoretical curves obtained from the FB theory.

Figure 5.12 shows experimental and theoretical values of the universal function $\Im(\omega^*)$ as well as results of measurements of ultrasonic absorption in a pure, admixture-free, propanol–water solution [137]. The good agreement between experimental and theoretical values of ultrasonic absorption demonstrates the applicability of the theory of dynamical scaling with regard to all phase transitions in a wide region of DCPs, up to the state of absolute homogeneity of a solution.

For a description of the acoustic properties of the 2-butanol–water + propanol system with three phase transitions, analogously to (5.22), we have

$$\frac{\alpha_s \lambda}{\alpha_s \lambda_{\max}} = \Im \left\{ \frac{3\omega \eta_p r_0^3}{k_B T} f(x/x_{c3}) \left[\frac{x_{0i} - x}{M x_{0i}} - \Delta t_{0i}^2 \right]^{2-\alpha} \left[\frac{(T - T_{c3})}{T_{c3}} \right]^{2-\alpha} \right\}, \tag{5.34}$$

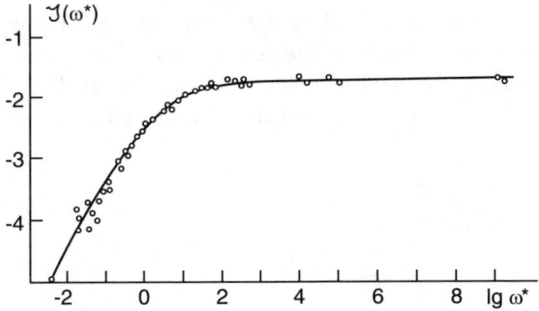

Fig. 5.12. Dependence of the scaling function on the reduced frequency

where $f(x/x_{c3})$ is a function characterizing the remoteness of the system from critical concentrations of the components.

The results of calculations using (5.34) are represented in Fig. 5.10b by solid lines. These are seen to agree with the data from measurements of ultrasonic absorption. As in the case of the first solution, all parameters of (5.34) were determined from independent measurements, including the value $r_0 f(x/x_{c3}) \approx 1.1$ Å found from optical measurements. One of the results of the acoustic experiment is connected with the constancy of the value of the critical absorption amplitude along the line T_c and in DCP, whereas in DCP the critical part of the heat capacity $C \to 0$ [63, 65]. This proves that the thermal character of acoustic relaxation at the critical point transforms into volumetric DCP, where dT_c/dP, characterizing the susceptibility of the system to pressure, tends to infinity.

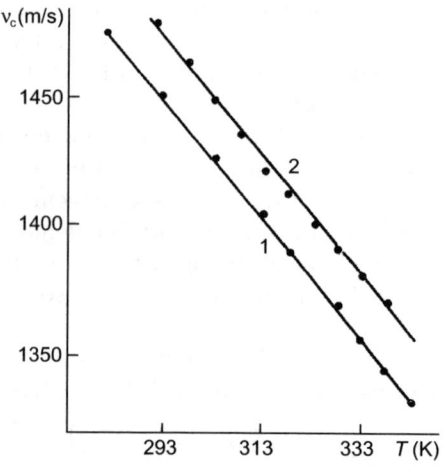

Fig. 5.13. Temperature dependence of ultrasound velocity near DCP for 2-butanol–water + propanol (1) and propanol–water + NaCl (2)

For the above-mentioned solutions, ultrasound velocity measurements were also performed at frequencies of about 30 kHz. It appeared that, within the range of measurement accuracy achieved (0.1), its temperature dependence at low frequencies assumes only a linear approximation (Fig. 5.13). Therefore, for reliable observation of ultrasound velocity anomalies in DCP, special measurements are needed to raise their accuracy at even lower frequencies.

6. Micellization as a Phase Transition

6.1 Conceptual Experiments

At the present time, physical studies of self-organization associated with the emergence of new structures occurring in many material systems is under extensive development [1, 78, 138, 139]. In particular, the problems associated with research into self-assembly phenomena of larger-than-molecular structures in liquid systems, involving various liquid-crystal states of organic molecules, are under active consideration. An important peculiarity of the above processes concerns the concentration of a dissolved substance in a system, which is a main physical state variable. They are known to be lyotropic systems.

In most cases, a prototype of liquid-crystal structures is a micellar state that occurs in surfactant solutions and is characterized by high dispersity and a finite number of particles in each structural unit, that is, in each micelle. Hence, investigation of the micellization process (i.e., structural units of lyotropic mesophases, whose forms and sizes determine their physical properties) makes it possible to elucidate the nature of the physical phenomena responsible for the liquid-crystal state in surfactant solutions. The main process is association of monomers, beginning with the emergence of dimers and ending with micellization.

Investigations into the mechanisms responsible for the association of monomers has had a long history. Adsorption of substances by a surface layer is known to be associated with low surface tension. In fact, accumulation of a substance in the interface predetermines that some parts of monomers are well soluble in one liquid and insoluble in other, e.g., hydrophylic and hydrophobic parts in the oil–water interface of two liquids. In short, the ability of monomers to form micelles comes about through the hydrophylic–lipophylic balance related to the degree of solubility.

Solubility is known [140] to depend exponentially on the work done by a molecule of a substance at the transition from its own phase into the solvent. It is assumed that the energy for introducing a molecule of the substance into water may be determined mainly through the work needed to form a cavity inside the water where a dissolved hydrocarbon molecule is to be located, and also by the energy needed to transfer it there. Thus, the hydrophobic effect is a manifestation of the interaction resulting from reassembly of hydrogen bonds and structural changes in the water.

The hydrophobic interaction is thought to be a major cause of micelle formation. However, as for monomers in water, it strongly depends on their environment in the finite state. It has been shown [141] that the total free energy of micellization, even regarding a molecular head, is less in absolute value than the change in the free energy on account of the hydrophobic part, under distribution in the hydrocarbon medium. The same conclusion also follows from calculations of the contribution made by the CH_2 group to a change in the free energy for chains of mean length [142]. This work examines the transfer of monomers from a water environment into a hydrocarbon liquid, where the change in the free energy was found to be equal to ~ 840 cal/mole, whereas upon transfer into the micelle it was only ~ 700 cal/mole.

Obviously, the micellization process is only partly caused by hydrophobicity. A lower value of the free energy of micelle formation compared with solvation in a hydrocarbon phase indicates a strained state of bonds and interlacing of hydrocarbon chains. Apparently, when packed into a micelle, hydrocarbon tails interact in such a way as to change their microstructures. Therefore, when analyzing mechanisms that provide thermodynamic stability of micelles, one should attend to the contribution of other, non-hydrophobic interactions.

Looking for possible mechanisms of micellization, it is useful to bear in mind that not all amphiphiles form micelles. The solubility of monomers differs significantly with regard to various amphiphilic molecules. For instance, dipalmitoylphosphatidylcholine in equilibrium does not form micelles [143]. Such regularity is also observed when hydrocarbon chains of molecules are weakly subjected to conformation changes. Hence, the ability to change conformation of molecules, and thus also the growth of microstructurization of a hydrocarbon nucleus, should be directly associated with micellization mechanisms. More penetrating investigations into the causes of thermodynamic stability of micelles are needed.

The thermodynamic stability of micelles is confirmed by the existence of micelles in overcooled micellar solution. In addition, micellization is an activating process characterized by a specific energy barrier. Once overcome, micellization is observed to prevail over small aggregate states of monomers in surfactant solutions. At small values of the surfactant concentration, however, micelles are in equilibrium with the solution of surfactants and, in essence, they are just microphases. Because of the absence of a macrophase, the Kraft point (T_K) cannot be interpreted as a triple point characterized by the existence of three phases, though there is a correlation between T_K and the effective melting temperature which characterizes the hydrocarbon part of a micelle.

As noted above, the packing of amphiphilic molecules in a micelle requires strong conformations in hydrocarbon chains. Therefore the Kraft temperature might somehow participate in the internal structural reassembly of the hydrocarbon nucleus of a micelle and the mechanism of its thermodynamic stability. At temperatures lower than T_K, the energy spent on chain conformation in micelle formation increases the thermodynamic potential. In this case, packing of monomers with lower curvature is preferable, as happens in the formation of a crystal phase. Thus, despite a wide range of studies on micelles, it is not yet possible to cover all the more obscure points of micelle formation.

In part, the problem of interpreting micelle formation in terms of phase equilibrium can be solved by the quasi-chemical approach. If one considers a molecular aggregation to be a chemical reaction, then a micelle is a natural product of the chemical reaction. Therefore, if in this case a micellar system can be viewed as a monophase, then the question of macro- or microphases is no longer relevant. The condition of aggregative equilibrium and the law of mass action for equilibrium states largely explain micelle formation phenomena. However, this assumes that the aggregation numbers are constant, i.e., independent of surfactant concentration. According to the quasi-chemical approach, micelles are formed by association of monomers with a molecular aggregate. The probability of this process rises with increasing surfactant concentration. In this case the activation energy of micellization can be generated by formation of intermediate molecular aggregates.

The limits of the quasi-chemical approach are clearly revealed when it comes to finding physical meanings for aggregation constants and justifying the assumption about constancy of aggregation number. By idealizing the overall picture and satisfactorily describing kinetic processes, this approach becomes complementary to the thermodynamic method. Thus, neither the latter nor the former approach can completely describe the main peculiarities of micelle formation, owing to the complexity of the investigated phenomena. It is thus of great importance to undertake comprehensive studies of the properties of micelles and external influences upon them.

One would think that the absence of a theory of micelles might limit their purposeful use. However, some of their important properties are now set on a firm basis. For instance, solubilization is undoubtedly related to hydrophobicity of the solubilized substance [144]. Swollen micelles can often be considered as small drops of an oil-type phase in a water environment.

Close to solubilization is the ability of micelles to serve as a reservoir for accumulating monomers in solution. When the concentration exceeds a critical value, the number of micelles increases and hence the growth of several thermodynamic parameters depending on this concentration is restrained.

Another important and exclusive peculiarity of micelles is their ability to change the rate of a chemical reaction. They can be divided into two types depending on the way they act: the reactivity is changed

- by localization of reagents inside a micelle, when a small volume of micelles actually raises the substrate concentration,
- by transfer of reagents into a micellar phase.

It is the second case that is assumed to be especially important in the theory of micelles. If understood, the concrete mechanisms that change the rate of chemical reactions in a micellar medium would considerably help us in explaining the nature of the micelles themselves.

Oxidation–reduction reactions in micellar solutions have proved to be another source of information on properties of micelles. Initiated by light, these reactions can be used for transformation of light energy into chemical energy. It is well known that the major difficulty in this case is the reversible character of photoprocesses. In a homogeneous solution, a direct transfer of electrons is usually accompanied by the fast reverse reaction. This gives impetus to a number of reactions which in turn result in thermal light scattering. Transformation of the light energy into a durable chemical form is successful only when the reverse reactions can be kept under control. Their rate can be reduced by a micelle system, for instance, via the regulated surface potential of a micelle. The first variant makes use of micelles as donors, i.e., a molecule that can be isolated in the visible region of light is directly built into the hydrogen section of a micelle. Here, in particular, the excited state of the molecule in a water solution is a better donor than the equilibrium state [145]. It should be noted that most of the organic sensitizers having small ionization energy are weakly soluble in water. In this situation, the reverse recombination reaction due to its high rate may come into play before the necessary channel of the chemical reaction is launched. If one dissolves a sensitizer in a micelle, it may become an excellent supplier of hydrated electrons, since the recombination reaction stops.

Another variant deals with the case when the laser irradiation energy is insufficient for formation of a free electron in a micelle. Then irradiation can only transform the donor molecule into the excited state. The further transfer of the electron into water can occur by tunneling. This is only possible if an excited state is reached, for which the hydrocarbon phase of the micelle must have a set of electronic states.

The dependence of the reaction rate constant of hydrated electrons with various acceptors on the potential difference between the hydrocarbon region of the micelle and the water phase [145] also evidences the presence of energy levels in the hydrocarbon nucleus of a micelle. When cation micelles have a high surface potential, the rate constant weakly depends on the affinity of acceptors to electrons. The permanence of the reaction rate constants of hydrated electrons with acceptors in cation micelles results from the long-time localization of an electron in the region of a micelle. The complex of electronic levels of the micelle is confirmed by comparative data on the degree of photoionization in pure liquids and ionic micelles. As shown in [146], the formation of hydrated electrons in a micelle is proportional to the square of the laser light intensity, leading to a non-linear dependence of the luminescence yields on the ionization intensity.

In surfactant molecules that contain unsaturated fragments of a non-conjugate cis-configuration, so-called loops can be formed in the geometry that significantly damage the compact crystal-like package of molecules. Such cis-unsaturated lipid acids in the nucleus of micelles give rise to dilution of the micro-environment.

Conformation of unsaturated fragments of hydrocarbon chains can markedly influence the process of peroxide oxidation caused by ionizing radiation [145]. Peroxide oxidation takes place if hydrocarbon chain loops form new electron states or introduce perturbations into the electronic spectrum. Therefore, an even greater change in the electronic spectrum of macromolecules may be observed when pre-micellar associates arise. Certainly, a strong conformational change in hydrocarbon tails upon packing molecules into a micelle cannot avoid influencing their electronic subsystems. On the other hand, it is known that in such systems the localization of free electrons in restricted regions of the changing volume [147] is able to stabilize this change thermodynamically. Meanwhile, the role of an electronic subsystem in the process of micelle formation has not yet been completely elucidated. In view of the above, it is assumed that, if electrons in a micelle can actively take part in photoionization, tunneling, oxidation, etc., then they can also make some contribution to the thermodynamic stability of micelles and hence to the theory of micelles.

6.2 Electronic Structure of Hydrocarbon Chains of Molecules

The basis of a micelle is a monomer containing a hydrocarbon chain. It is known that such a chain cannot avoid influencing a large number of saturated and unsaturated bonds with neighboring atoms. Although the electron structure of systems with unsaturated bonds has been studied in some depth [148], the same cannot be said for macromolecules with σ-bonds. For the investigation of macromolecular systems, the methods of the self-consistent field and of molecular orbitals are used. These begin by assuming that σ-bonds form deeply-lying filled zones. The properties of the group of atoms are characterized only by the number of π-electrons. The methods of calculation used for molecules with unsaturated bonds are also proved to be appropriate, after some correction, for systems with σ-bonds [149]. Though these methods cannot provide one with exact values of the object parameters, they allow us to obtain a qualitative description of the relevant properties.

To describe a hydrocarbon chain in [150], the Hubbard monomer Hamiltonian [151] was used:

6. Micellization as a Phase Transition

$$H = \vartheta \sum_{i,\sigma} \hat{B}^+_{i,\sigma}(\hat{B}_{i+1,\sigma} + \hat{B}_{i-1,\sigma}) + \zeta \sum_{i,\sigma} \hat{n}_{i,\sigma}\hat{n}_{i,-\sigma} ,\qquad(6.1)$$

where $\vartheta < 0$ is a resonance integral, $\zeta > 0$ is a parameter of the Coulomb interaction of electrons in an atom, and $\hat{B}^+_{i,\sigma}$, $\hat{B}_{i,\sigma}$ are the creation and annihilation operators for the ith atom with spin σ.

Using the approximate Hartree–Fock equation, the following has been obtained:

$$E_l B^l_{i,\sigma} = \alpha_0 B^l_{i,\sigma} + \vartheta(B^l_{i+1,\sigma} + B^l_{i-1,\sigma}) + \zeta \hat{n}_{i,-\sigma} B^l_{i,\sigma} .\qquad(6.2)$$

Its solution has the form

$$E_l = \alpha_0 + 0.5\zeta \pm \left[(\zeta\delta)^2 + 4\vartheta^2 \cos^2(\pi l/N)\right]^{1/2} ,\qquad(6.3)$$

where N is the number of atoms in a chain, $B^l_{i,\sigma}$ is the expansion coefficient of the molecular orbital Ψ_σ,

$$\hat{n}_{i,\sigma} = \sum_l |B^l_{i,\sigma}|^2 \hat{n}_{l\sigma}$$

is the average over the sought state, α_0 is the Coulomb integral, and δ is a parameter.

If $N = 2p+1$ is odd then, in the ground state, the levels which correspond to the minus sign of the radical and for which $-p \le l \le p$, are filled. The plus sign corresponds to excited states. Depending on δ, there are two solutions. The first, $\delta = 0$, is in agreement with the results obtained by the Hartree–Fock method when

$$\Delta A_l = 2\vartheta \cos \frac{\pi l}{N} .\qquad(6.4)$$

When $\delta \ne 0$, the energy of the ground state of a hydrocarbon chain of the polyene type proves to be lower than that when $\delta = 0$. The ground state of the system when $\delta \ne 0$ is separated from the excited states by a gap

$$\Delta A = 2\zeta\delta .$$

As shown in [150], the solution with $\delta = 0$ is unstable and can be transformed into a stable one with a small perturbation giving rise to spin polarization of the electron shells of the chain. For instance, an electron added into the chain may constitute such a perturbation. In accordance with the calculation, $E(\delta) < E(0)$. Hence, the energy gap proves to be different for even and odd numbers of hydrogen atoms in the chain.

Consequently, deformations of interatomic bonds, changes in conformation of the chain, and the joining on of other atoms are able to generate local states, thereby bringing the hydrocarbon chain closer to a metallic state. For instance, the presence of local states explains the difference in behavior of the first electron transition frequency in symmetrical dye-stuffs and in polyene chains. Introduction of nitrogen atoms into the chain seems to bring about local states near zone boundaries. It results in electron transitions from the local state to a free zone which, under additional conditions, can narrow the energy gap. This is very clearly observed if the ideal chain periodicity is destroyed when the break causes splitting of the local levels off zones of resolved states. The levels move into regions either higher or lower than both of the resolved zones. Assuming that either alteration of bond lengths or correlative interaction of electrons leads to formation of the gap, the conditions for local states to arise under the simplest perturbations of hydrocarbon chains were created in [150] using different methods. According to these models, perturbation resulting from a change $\Delta \alpha$ in the Coulomb integral of the nth even atom splits off a local level in the forbidden zone only if $|\Delta \alpha| \geq 4\zeta$, whereas alterations in an odd atom already lead to splitting under an infinitesimally small perturbation. This fact deserves special attention because in this case the oscillatory dependence of physical properties of molecular associates on the length of hydrocarbon tails of surfactant molecules is already manifested on the electron level.

6.3 Fluctuon Model of Micellization

Let us consider a water solution of surfactant molecules in equilibrium at minimal free energy. When concentrations of monomers are small, their distribu-

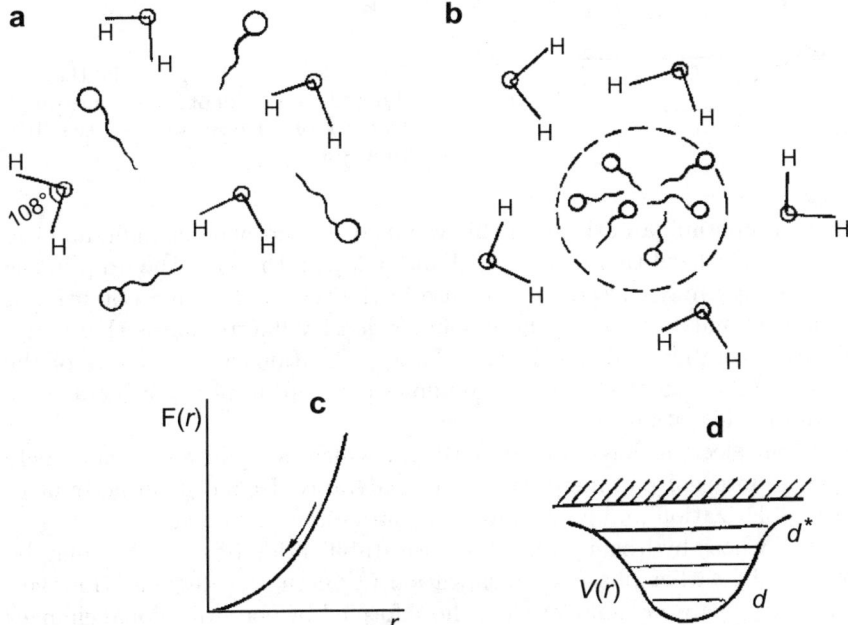

Fig. 6.1. a–d. Formation of fluctuon states of electrons in water solutions of surfactants

tion may be considered to be homogeneous (Fig. 6.1a). As the concentration increases, the probability of random accumulation of a group of monomers in a closed region of the solution gets higher (Fig. 6.1b). However, such fluctuon formation of monomers of high density raises the thermodynamic potential of a system (Fig. 6.1c), which is energetically unfavorable, and the fluctuation resolves. As is known from the theory of disordered media [156], a non-equilibrium increase in the local density of a system is accompanied by formation of local energy levels beneath the bottom of the conductivity zone.

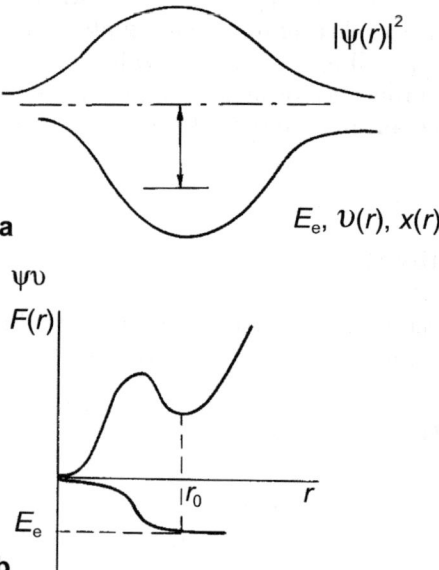

Fig. 6.2. a,b. Change in the thermodynamic potential of a system through transfer of an electron into the fluctuon state

If the potential gap arising in this way is deep and wide enough, an electron can be localized there (Figs. 6.1d and 6.2a). In this case the drop in the electron energy may exceed the increase in the thermodynamic potential at fluctuon accumulation. Then the monomer density fluctuation is thermodynamically favorable, and the density change is stationary. The curve of the thermodynamic potential (Fig. 6.2b) shows the appearance of a metastable minimum in this situation.

Once an electron has entered the trap, which is achieved via an easily changeable internal parameter, it can considerably deepen it on account of electron polarization and reorientation of surfactant monomers.

Water, whose hydrogen atoms increase stabilization of electrons, may be present in the earlier phases of localization in fluctuation regions. However, such potential gaps are smaller than those formed by conformational changes in hydrocarbon sections of monomers at a later phase of auto-localization. Packing of hydrophobic sections inside the fluctuation, and considerable

6.3 Fluctuon Model of Micellization

bending and fracturing of hydrocarbon chains significantly deepen the potential gap and add energy levels there. The types of hydrocarbon tails of the surfactant molecules determine whether some or all the water should be removed from the center of such a hydrocarbon cluster.

Hence, after auto-localization of the electron, the fluctuon accumulation of monomers soon turns into a micelle. In this case the electron acts as a micelle 'builder'. The existence of local electron states upon accumulation of monomers and conformational changes in hydrocarbon chains has been convincingly demonstrated above. However, one question remains: where does the electron come from?

If there are conjugate bonds in the hydrocarbon section of the monomers, the presence of a weakly associated, effectively delocalized π-electron would predetermine its localization under conformational changes in the hydrocarbon chain. However, the hydrophobic section of monomers is usually formed by saturated σ-bonds. Apparently, in this case the donor is a benzene ring, from which an electron can transfer into a local state in the fluctuon accumulation.

Another source of electrons can be donor admixtures in water which are able to return an electron to the water. This results in hydrated electrons which are held in small traps made by water molecules [157]. Experiments with laser irradiation of water have revealed the possible participation of hydrated electrons. Hydrated electrons may also be detected by EPR signals. Moreover, in both cases luminescence [158] confirms the active participation of monomers and micelles in relaxing electron excitations in the solution.

Thus, for further development of the fluctuon model, one needs to understand whether self-crossing of hydrocarbon tails of surfactant molecules with saturated electron bonds is able to decrease the energy of the system.

Hydrocarbon tails of monomers can be identified with an equally bonded one-dimensional chain. The wave function of the electron in this chain can be written in the form

$$\mid \Phi_k(x,y,z) \mid^2 = \mid \Phi_k(x) \mid^2 \delta(y)\delta(z) \:, \tag{6.5}$$

where $\Phi_k(x)$ is the electron wave function in the one-dimensional chain along its length (x axis), and $\delta(y)$, $\delta(z)$ are delta functions.

As explained above, the electron structure of this chain was calculated using various approaches. To answer the question raised above, we consider a group of models employing the Green function method and Huckel's method. To characterize the two approaches, we modify the original structure as follows: let a chain have one self-crossing such that the distance between two crossing nodes will be equal to the chain period.

6.4 Green Function Method

In the single-electron approach, we consider the Green function G satisfying the operator equation [159]

$$G = G^0 + G^0 \hat{v} G, \tag{6.6}$$

where G^0 is the Green function of a straight-line chain, v is a perturbation formed on account of self-crossing of the chain, and G is the Green function of the chain with the form

$$G^0(x, x^*, \omega) = -\frac{1}{N} \sum_k \Phi_k(x) \Phi_k^*(x^*) G^0(k, \omega), \tag{6.7}$$

where

$$G^0(k, \omega) = \frac{1}{\omega - A \cos(ka)}.$$

Let us form a loop on the macromolecule (Fig. 6.3a). For this special loop conformation, only the interaction of nodes p and q is new, and this is written as

$$\hat{v} = v \delta(p - q) + v^* \delta(p - q), \tag{6.8}$$

where $\delta(x)$ is the Dirac delta function and $\hat{v} = v a_p^+ a_p + v^* a_q^+ a_q$ (i.e., we use second quantization).

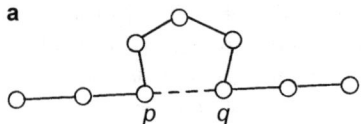

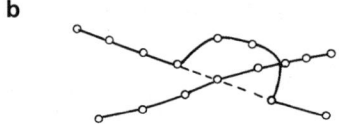

Fig. 6.3. a,b. Possible conformation changes in hydrocarbon radicals of surfactant molecules

The Dyson equation for the excited system + polymer with a loop is then (6.6). Using a standard method, Fourier expanding both the potential $\hat{v}$ and the Green functions $G(x, x^*, \omega)$ and $G^0(x, x^*, \omega)$ and then summing, we get [159]

$$1 - \frac{2|A|}{(\omega^2 - A^2)^{1/2}} + \frac{A^2}{\omega^2 - A^2} \left[\frac{A^2}{\omega^2 - A^2} - \frac{A}{\omega - (\omega^2 - A^2)^{1/2}} \right]^m = 0, \tag{6.9}$$

where $m = |p - q|$. The solution of this equation in the energy region $\omega > |A|$, i.e., lower than the band of continuous energy values, proves to depend on the distance between nodes $|p - q|$.

Thus, at small distances $|p - q|$, the local level E is actually split off the zone that makes the loop conformation more favorable than in the linear case. However, the equations describing such complex systems as conformations of hydrocarbon radicals are very difficult ones and it is impossible to obtain exact solutions. One needs to develop effective approximation methods for solving quantum mechanical problems, which can be used to describe the main properties of complex atom combinations without going into extremely time-consuming calculations. In order to understand the essence of the investigated subjects more easily, it is useful to introduce a simplified model of the system. These simplifications do not distort the essence of the phenomena under investigation, and at the same time they are so effective that they allow us to extract much more information without unwieldy calculations. One such method is the model suggested by Huckel, the theory of molecular orbitals.

6.5 Huckel's Method of Molecular Orbitals

Let us consider a non-closed surfactant molecule consisting of 8 hydrocarbon links [160]. The hydrophobic tail can be considered separately because the interaction of the hydrophilic head with the nucleus of the micelle is much weaker than the hydrophobic interaction [161].

In the Huckel approach to obtain the energy spectrum of a system, the Huckel determinant is written

$$\begin{vmatrix} E-\alpha & \vartheta & 0 & 0 & 0 & 0 & 0 & 0 \\ \vartheta & E-\alpha & \vartheta & 0 & 0 & 0 & 0 & 0 \\ 0 & \vartheta & E-\alpha & \vartheta & 0 & 0 & 0 & 0 \\ 0 & 0 & \vartheta & E-\alpha & \vartheta & 0 & 0 & 0 \\ 0 & 0 & 0 & \vartheta & E-\alpha & \vartheta & 0 & 0 \\ 0 & 0 & 0 & 0 & \vartheta & E-\alpha & \vartheta & 0 \\ 0 & 0 & 0 & 0 & 0 & \vartheta & E-\alpha & \vartheta \\ 0 & 0 & 0 & 0 & 0 & 0 & \vartheta & E-\alpha \end{vmatrix},$$

where E is the system energy, α is the ionization potential of an isolated atom, and ϑ is the exchange integral of a linear molecule. Let us introduce the notation $z = (E-\alpha)/\vartheta$. Then the above determinant takes the form

$$\begin{vmatrix} z & 1 & 0 & 0 & 0 & 0 & 0 & 0 \\ 1 & z & 1 & 0 & 0 & 0 & 0 & 0 \\ 0 & 1 & z & 1 & 0 & 0 & 0 & 0 \\ 0 & 0 & 1 & z & 1 & 0 & 0 & 0 \\ 0 & 0 & 0 & 1 & z & 1 & 0 & 0 \\ 0 & 0 & 0 & 0 & 1 & z & 1 & 0 \\ 0 & 0 & 0 & 0 & 0 & 1 & z & 1 \\ 0 & 0 & 0 & 0 & 0 & 0 & 1 & z \end{vmatrix}.$$

6. Micellization as a Phase Transition

If the macromolecule has bends in nodes p and q (Fig. 6.3a), then the determinant is equal to

$$\begin{vmatrix} z & 1 & 0 & 0 & 0 & 0 & 0 & 0 \\ 1 & z & 1 & 0 & 0 & 0 & 0 & 0 \\ 0 & 1 & z & 1 & 0 & 0 & \varepsilon & 0 \\ 0 & 0 & 1 & z & 1 & 0 & 0 & 0 \\ 0 & 0 & 0 & 1 & z & 1 & 0 & 0 \\ 0 & 0 & \varepsilon & 0 & 1 & z & 1 & 0 \\ 0 & 0 & 0 & 0 & 0 & 1 & z & 1 \\ 0 & 0 & 0 & 0 & 0 & 0 & 1 & z \end{vmatrix}.$$

Then elements with indices 2.7, 6.3 are seen to equal not zero, but the value $\varepsilon = \vartheta'/\vartheta \ll 1$, where ϑ' is the resonance integral of nodes p and q of the molecule. Expanding the determinant, we find that

$$\det A = \det A_0 + 2\varepsilon(z^5 - 3z^3 + z) + O(\varepsilon^2), \tag{6.10}$$

where

$$\det A_0 = z^8 - 7z^6 + 15z^4 - 10z^2 + 1 \tag{6.11}$$

is the expansion of the determinant for the molecule without crossing. Neglecting the second order terms $O(\varepsilon^2)$, solutions can be sought in the form

$$z = z_0 + \delta, \tag{6.12}$$

where z_0 are the roots of (6.11).

Substituting (6.12) into (6.10), we expand the latter as a power series in δ/z_0. In the linear approximation for δ, we obtain

$$\delta = \frac{-2\varepsilon(z_0^5 - 3z_0^3 + z_0)}{8z_0^7 - 42z_0^5 + 60z_0^3 - 20z_0 + 1 + 2\varepsilon(5z_0^4 - 9z_0^2 + 1)}.$$

This is used to evaluate the distortion of the local level E^* from the zone, and in this case $E^* = E_0 + \delta\vartheta$. For two surfactant monomers forming a loop and crossing (Fig. 6.3b) the value of E can also be found. Expansion of the 16×16 determinant gives the relation

$$\det A_{16} = (\det A_0)^2 + 2\varepsilon(z^8 - 6z^6 + 5z^4 - 6z^2) + O(\varepsilon^2).$$

Solution of the corresponding equation shows that new electron states appear with negative energy sign.

As seen from the calculations (Table 6.1), when the perturbation value as a result of conformational reformation of hydrocarbon radicals is 100 times less than the exchange interaction ϑ, i.e., $\varepsilon = 0.01$, at $\alpha = -11$ eV, $\vartheta = -2.3$ eV [149], the local level actually splits off the zone and the difference of these levels is about 0.02–0.05 eV. The filling of these states stabilizes the above-mentioned conformation states of monomers, and hence also the micelle.

This method is simple to use and quite adequate for any conformation with an arbitrary number of loops. It can be used for a complete study of the

Table 6.1. Energy values for a hydrocarbon radical, unperturbed (E) and perturbed (E^*) by a conformation change

z_0	δ	E	E^*
1.00	0.00420	−13.000	−13.028
−1.00	−0.00420	−8.200	−8.201
1.88	0.00129	−16.262	−16.266
−1.88	−0.00129	−5.737	−5.733
1.53	0.00071	−15.288	−15.290
−1.53	−0.00071	−6.712	−6.709
0.35	0.03449	−11.972	−11.875
−0.35	−0.03449	−10.028	−10.051

genesis of electron states in loop formation. In particular, it helps to evaluate a change in the depth A of a potential well resulting from the conformational alterations in hydrocarbon radicals as a function of the number N of atoms in a chain (Fig. 6.4).

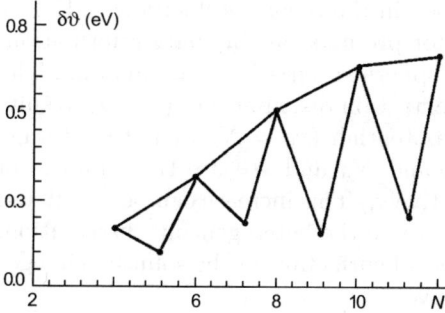

Fig. 6.4. Dependence of the depth of a potential well on the number N of atoms in a chain

Calculations of the electron states for a linear molecule and a molecule with a loop lead to the expressions

$$\det A_N = z \det A_{N-1} - \det A_{N-2},$$

$$\det A_N^* = z \det A_{N-1} - \det A_{N-2} + \Omega(N) 2\varepsilon + O(\varepsilon^2),$$

where

$$\Omega(N) = \begin{cases} 1 & N = 2n+1, \\ -1 & N = 2n+2. \end{cases}$$

84   6. Micellization as a Phase Transition

According to the calculations, while transferring from an odd number of atoms to an even one in the spectrum of electronic states of hydrocarbon radicals, $\delta\vartheta$ is observed to oscillate as N increases:

N	6	7	8	9	10	11	12
$\delta\vartheta$	0.22	0.12	0.41	0.06	0.64	0.06	0.67

This is due to alternate filling in of the highest occupied molecular orbital by either one or two electrons. The experimental measurement of critical micellization concentration (CMC) [161] for surfactant molecules with odd and even numbers of hydrocarbon atoms in a chain has shown that the hydrocarbon nucleus of a micelle has solid-like properties.

6.6 Critical Micellization Concentration

When investigating a fluctuation state of electrons, the mobility of atoms is considered to be relatively high, provided that sources of the field redistribute themselves as a result of interaction with an auto-localized electron. In view of this, we will further investigate the case where the relaxation time τ_0 of the medium in a small region of the fluctuation potential well is smaller than the lifetime τ of an electron in this well, i.e., $\tau_0 \ll \tau$.

Following the methodology developed in the theory of fluctuons [147], let us study the main equations of state for pre-micellar fluctuation formation. For simplicity, we limit ourselves to a spherical particle approximation. The appearance of the heterogeneous concentration distribution $x(r) - x_0$ results in the growth of the thermodynamic potential. ($x_0 = N/N_A$ is the average concentration, where $N = N_A + N_B$, and N_A and N_B are the numbers of atoms of surfactant and solvent, respectively.) This increase can be calculated as the minimal work $Rx(r)$ needed to create the heterogeneous distribution. We assume that the concentration $X(n)$ of surfactant in the solution changes gradually. From this assumption we have

$$R = \int \left[\varphi[x(r)] - \varphi(x_0) - \frac{\partial \varphi(x_0)}{\partial x_0}[x(r) - x_0] + \frac{1}{2}B_1(\nabla x)^2 \right] dr ,$$

where $\varphi[x(r)]$ is the density of the thermodynamic potential of aqueous surfactant solution, and B_1 is a coefficient related to the heterogeneous accumulation of the concentration.

As shown above, disorder in a system of surfactant molecules leads to a splitting of the electron levels of these molecules. Moreover, their accumulation itself promotes blurring in the solution conductivity zone. Let us choose the boundary of the conductivity zone as a zero reading with respect to the electron energy and, for the sake of simplicity, assume that in the region of changeable surfactant concentration, this energy is linear in the concentration:

$$\vartheta(r) = A\,|x(r) - x_0|\;.$$

If $\vartheta(r) < 0$ and the resulting potential well is wide enough, then the electron has a negative energy $E_e < 0$ when localized in it. When the electron is transferred from a small trap or a highly deformed hydrocarbon tail to the concentration fluctuation, the change in the thermodynamic potential of the system (solution of surfactant molecules + electron) is as much as $R + E_e$. As shown in [143], the electron energy $R + E_e$ at the given value of the concentration $E[x(r)]$ in the adiabatic approach can be determined as a minimum with respect to normalized functions $x(r) - x_0$, $\Psi(r)$ of the functional $E[x(r), \Psi(r)]$ representing the quantum mechanical average of the energy operator. The stationary distribution of the concentration will correspond to a minimum of the functional $R[x(r)] + E[x(r)]$ with respect to $x(r)$. Hence, a change in the concentration can be determined by the minimum condition with respect to Ψ and $x(r)$ of the functional

$$\Im = \frac{\hbar^2}{2m}\int |\nabla\Psi|^2\,dr + \int A[x(r) - x_0]\,|\Psi|^2\,dr + R[x(r)]\;. \tag{6.13}$$

This functional is more convenient for minimization with respect to $x(r)$ at the given $\Psi(r)$. As a result, we obtain

$$\frac{\partial\varphi[x(r)]}{\partial x(r)} - \frac{\partial\varphi(x_0)}{\partial x_0} + A\,|\Psi(r)|^2 = 0\;.$$

This equation describes the distribution of the concentration $x[\Psi(r)]$ in the effective potential $A\,|\Psi(r)|^2$ generated by an auto-localized electron. The term $B_1(\nabla x)^2$ is omitted when working on the microscopic level. After substituting $x[\Psi(r)]$ corresponding to the minimum of $\Im$ with respect to x, we obtain an expression for the functional $\Im[\Psi]$ depending on a single variable. A spherically symmetrical wave function corresponds to the ground state of the associate.

For an ideal solution, one can neglect the interaction energy of molecules of different types. In the macroscopic approach, the minimal work for creation of the heterogeneous distribution $x(r) - x_0$ can be written in the form

$$R = \frac{k_B T}{V}\int\left(x\ln x - x\frac{\bar\sigma S_1}{k_B T} + \frac{\bar\sigma S}{k_B T}\right)dr\;, \tag{6.14}$$

where $\bar\sigma$ is the average surface tension, S_1 is the surface area of a micelle per monomer, S is the total surface area of a micelle, and V is its volume.

A change in the thermodynamic potential of a micelle depending on the localization of the electron can be determined by the minimum condition on the functional $\Im[x(r), \Psi(r)]$ given in (6.13). By minimizing $\Im$ with respect to $x(r)$, we obtain an obvious expression for the heterogeneous distribution $x(r)$, viz.,

$$x(r) = \Theta(\sigma,T)\exp\left(-\frac{AV}{k_BT}|\Psi|^2\right), \quad \Theta(\sigma,T) = \exp\left(\frac{\bar\sigma S_1}{k_BT}-1\right). \tag{6.15}$$

The minimum of the functional $\Im(\Psi)$ can be determined by a variational method. By making use of the simplest approximation for Ψ,

$$\Psi(r) = \left(\frac{2\alpha^*}{\pi}\right)^{3/4}\exp(-\alpha^* r^2), \tag{6.16}$$

and substituting (6.15) and (6.16) into (6.14), we can write the function $\Im$ in terms of a variable a, inversely proportional to the effective volume of a micelle:

$$\Im(a) = -Ax_0 + \frac{A}{\pi^{3/2}}\left[\left(\frac{k_BTa}{A}\right)^{2/3}F + 4\frac{\Theta(\sigma,T)}{a}\left(f(a) + \frac{\sqrt\pi}{4}(b-1)\right)\right], \tag{6.17}$$

where

$$f(a) = \int_0^a (1-r)\ln^{1/2}\frac{a}{r}dr,$$

$$a = \frac{AV}{k_BT}\left(\frac{2\alpha^*}{\pi}\right)^{3/2}, \quad F = \frac{\pi^{3/2}\hbar^2}{8mAV^{2/3}}, \quad b = \frac{\bar\sigma S}{k_BT\Theta}.$$

This gives

$$f(a) = \begin{cases} \dfrac{\sqrt\pi}{2}a\left(1 - \dfrac{a}{\sqrt{2^3}} + \dfrac{a^2}{\sqrt{3^3}}\right), & \text{at } a \sim 1, \\ f(a) = \sqrt{\ln a}[1 - \exp(-a)], & a \gg 1. \end{cases}$$

From the condition $\Im(a) = 0$, one can find the value of $a = a_0$ corresponding to the minimum of the function, as well as the effective number n of monomers in a micelle, and the change in the thermodynamic potential of the system $\Delta\Phi$:

$$n = \frac{1}{\gamma_1(a_0)}\left(\frac{AF}{k_BT\Theta}\right)^{3/5},$$

$$\Delta\Phi = -Ax_0 + n\gamma_2(a_0)\frac{F^{3/5}}{\pi^{3/2}}\left(\frac{\Theta k_BT}{A}\right)^{2/5}, \tag{6.18}$$

where

$$\gamma_1(a) = \left[\frac{3}{4}[f_1(a) - af_1(a)]\right]^{3/5}, \quad \gamma_2 = \gamma_1^{2/3} + \frac{f_1}{\gamma_1},$$

and

$$f_1 = f(a) + \frac{\sqrt{\pi}}{4}(b-1).$$

Assuming the micelle surface tension of sodium dodecylsulphate to be as much as $\sigma \sim 5$–12 mN/m and the radius of a micelle $R_m \sim 25$ Å, we obtain the temperature dependence of the number n of monomers in a micelle (Fig. 6.5).

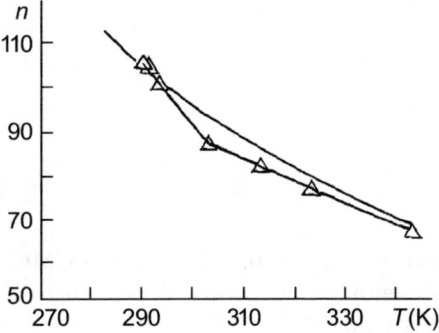

Fig. 6.5. Temperature dependence curves of the number of monomers in a micelle

Apparently, micelles exist when the electron energy $E < 0$. This corresponds to the region where $\Im(a) < 0$. However, when the temperature T increases, the minimal value of the functional $\Im(a)$ increases, and at some temperature $T = T^*$, the minimum of the functional $\Im(a)$ is positive and micellization does not occur. Calculations show that $n = 20$ at $A = 100 k_B T$ and $n = 25$ at $A = 300 k_B T$, i.e., as the temperature lowers, the number of monomers in a micelle grows. In this case the critical micellization concentration (CMC) is determined by the following function of σ and T [162]:

$$x_{\text{cmc}}^{\text{e}} = \left(\frac{x_V^* \Lambda^3}{V^{*3/2}}\right)^{1/V^*} \Lambda^{-3} \exp\frac{\Im(a_0)}{k_B T}, \qquad (6.19)$$

where x_V^* is a minimal, experimentally observable concentration of micelles, and n corresponds to the most probable size of a micelle. The superscript on $x_{\text{cmc}}^{\text{e}}$ indicates that electron localization is taken into account.

At $v^* = 50$, $x_V^* \sim 10^{17} \text{cm}^{-3}$ in the approach $\sigma S/k_B T < 1$, $f(a_0/a) \sim 1$, the form of the equation (6.19) greatly simplifies:

$$\ln(x_{\text{cmc}}^{\text{e}} \Lambda^3) \approx \frac{A}{k_B T} x_0 - \frac{\sigma S}{k_B T}. \qquad (6.20)$$

Since the CMC x_{cmc}^0 without taking into account the electron localization is determined by the expression [163]

$$\ln(x_{\text{cmc}}^0 \Lambda^3) \approx -\frac{\sigma S}{k_B T}, \qquad (6.21)$$

equation (6.19) can be written in the form

$$x_{\text{cmc}}^{e} = x_{\text{cmc}}^{0} \exp\left(-\frac{|A|}{k_B T} x_0\right). \tag{6.22}$$

Thus, the role of an electron subsystem is reduced to a shift in the values of x_{cmc} to a region of smaller magnitudes. Table 6.2 shows the ratios $x_{\text{cmc}}^{e}/x_{\text{cmc}}^{0}$ calculated by (6.20) and (6.21).

Table 6.2. Values of $x_{\text{cmc}}^{e}/x_{\text{cmc}}^{0}$ for homologues of potassium carbonates

a [Å]	σ [mN/m]	bn	$x_{\text{cmc}}^{e}/x_{\text{cmc}}^{0}$
272	9.7	25	0.60
380	5.9	31	0.77
489	4.2	34	0.83

At $A = 100 k_B T$, we obtain a correction $x_{\text{cmc}}^{e} = 0.6 x_{\text{cmc}}^{0}$ for the CMC. Comparing the theoretical and experimental temperature dependence of x_{cmc}^{e}, we may judge the reliability of this model. In the experiments, the CMC significantly decreases when electrolytes are introduced into the solution. The CMC of sodium and potassium dodecylsulfonate mixtures decreases by 40% [164] if 0.25 m.f. of $CaCl_2$ is added and it lowers by 27% if there is 0.02 m.f. of NaCl in water solutions of n-octylbenzolsulfonate [165]. This decrease in CMC with increasing concentration of donor admixtures indirectly corroborates the model.

6.7 Micellization as a Phase Transition of Finite Type

Phase transitions of the first order are known to be primarily determined by latent heat, volume change and hysteresis [166–168]. Micellization of the solution containing surfactant molecules is equivalent to the appearance of new excitation modes in the system. The latter result in a discrepancy between the heat capacities of two phases and discontinuous behavior of heat capacity in the system. Moreover, experimental data unambiguously show the existence of a positive volumetric effect whose value decreases as the temperature increases.

This would seem to be sufficient justification for the statement that the transition from a homogeneous to a micellar solution develops as a phase transition of the first order. However, despite the applicability of such terms as boundary surface and surface tension for micelles, micelles themselves cannot pretend to be the nuclei of a new phase because they do not have a macroscopic analog [55]. With increasing surfactant concentration, they remain in equilibrium, whilst changing in size. In essence, they form an independent phase with a strongly curved surface [143]. If one considers the state

6.7 Micellization as a Phase Transition of Finite Type

of monomers in a micelle to be more ordered, micellization can be thought as a transition between macroscopic phases from one solution into another with a change in symmetry, i.e., a phase transition of the second order. This point of view is confirmed by the absence of hysteresis (or only a weak presence), and an abrupt change in the physicochemical properties of the solution at CMC. Thus, it would be of interest to discover what type of phase transition is realized in the micellization process.

Micellization of surfactants in a water solution actually corresponds to the appearance of new excitation modes in the system. This should be accompanied by a change in the heat capacity, which can be evaluated by the relation

$$\Delta C \approx \frac{N_f T^*}{\delta T} \frac{d\Delta\Phi}{dT} = k_B N_f \frac{2}{5\pi^{3/2}} \gamma_2(a_0)\Theta^{2/5} \left(\frac{AF}{k_B T}\right)^{3/5} \frac{T^*}{\delta T}$$
$$= \frac{k_B N_f}{\ln(N_e\zeta/N_f)} \frac{2}{5\pi^{3/2}} \gamma_1(a_0)\gamma_2(a_0)n\Theta^2 , \tag{6.23}$$

where N_f is the number of forming micelles, δT is the temperature interval for transition of most monomers into a micellar state, and T is the temperature of the fixed concentration of monomers.

Since the micellar state is separated by a potential barrier from donor electron states, both micelles and donor electrons may exist simultaneously in the system. The number of micelles can be determined by

$$N_f = N_c\zeta \exp\left(-\frac{\Delta\Phi}{k_B T}\right) , \tag{6.24}$$

where ζ is the probability of an electron transition from the donor state to a micelle.

One can evaluate δT by varying the relation (6.24) near T, taking into account $\partial\Delta\Phi/\partial T$, and supposing N_e/N_f to vary from 0.1 to 1:

$$\delta T \approx \frac{5\pi^{3/2}}{2} \frac{T^*}{\gamma_1(a_0)\gamma_2(a_0)n\Theta^2} \ln\left(\frac{N_e}{N_f}\zeta\right) .$$

The calculations show that for sodium dodecylsulphate with $T^* \sim 313$ K, we find $\delta T \approx 1$–3 K [141], and the smeared peak of the heat capacity is about 200 J/(mol/g) at the concentration 8×10^{-3} mol/l. Thus, according to this model the existence of the heat capacity peak in a narrow range of temperatures implies that micellization is a smeared phase transition of the first order, as suggested by the results of [165].

The width of the potential well increases with an increase in the number of methylene groups in a hydrophobic tail, and so does the coefficient A characterizing the potential well. As seen from (6.23), the heat capacity peak rises with an increase in A, as observed experimentally [141].

6.8 Phase Transitions at Micellization in Solutions of Ionic Molecules

For the development of the micelle theory, experimental results embodying the macroscopic approach to this process are of great importance. Since micellization occurs in very dilute solutions and over a narrow concentration range, one needs highly sensitive and precise experimental methods. The acoustic method gives valuable information on the phenomena near a critical point. Recently developed ultrasound interferometers make it possible to carry out precise measurements of the sound propagation velocity v_c and, together with rheological data, to register small changes in the structure of solutions [78, 171].

The ionic surfactant sodium n-octylbenzenesulphonate (OBS) was chosen as the research subject in [161, 172, 173] because of its relatively large CMC. Measurements of density and viscosity of water solutions of OBS have shown that, in the considered temperature range, there is a break corresponding to the CMC in the property–concentration curve. The dependence of the density and viscosity of solutions on surfactant concentration both before and after CMC is well approximated by linear equations. These data can be used to calculate the thermodynamic parameters of micellization in water solutions of OBS, as well as some structural characteristics of OBS micelles (Table 6.3).

The analysis revealed that the behavior of OBS solutions is close to the behavior of ideal solutions both before and after CMC. Micellization occurs with a small thermal and positive volumetric effect. OBS micelles are statistical formations with a 'flocculent' hydrocarbon nucleus which have a chaotic arrangement of chains and are surrounded by a solvate membrane whose dimensions are defined mainly by the micelle surface charge. According to the results of measurements at 2 MHz using the resonance method and 10.9984 MHz using an interferometer with a variable base [170, 173–176], the speed of sound reaches a minimum at CMC (Fig. 6.6). Measurements at 10.9984 MHz made it possible to define a narrow range of changes in the speed of sound as a function of the concentration in the vicinity of the CMC (sections I, II, III in Fig. 6.6).

In the initial region, the nonlinearity of the speed of sound was established as being due to peculiarities in the variation of the solution density before and after CMC. The dependence of density on OBS concentration is approximated by the relations

$$\text{before CMC} \quad \rho = 998.73 + 0.56\varepsilon^{1/2}, \quad \text{at } \varepsilon \leq 0.0302, \quad (6.25)$$

$$\text{after CMC} \quad \rho = 998.78 + 0.48\varepsilon^{1/2}, \quad \text{at } \varepsilon \leq 0.0277, \quad (6.26)$$

where $\varepsilon = |x - x_c|/x_c$ is the remoteness from CMC.

The character of changes in the speed of sound in the section near CMC is related to peculiarities in the solution densities and in the adiabatic compressibility β_S near CMC. Far from CMC, β_S is described by

Table 6.3. Thermodynamic and structural characteristics of micellar aqueous solutions of OBS. ΔG_m, ΔH_m, ΔS_m, ΔV_m are changes in the free energy, enthalpy, entropy and partial molar volume of OBS, respectively, n the number of associates, r_h the hydrodynamic radius, D_m the coefficient of diffusion of OBS micelles, η_el the contribution of the electroviscous effect, and h^* the number of associated water molecules in a hydrodynamic particle

T [K]	x_cmc $\left[\frac{\mathrm{mol}}{\mathrm{m}^3}\right]$	ΔG_m [kJ/mol]	ΔH_m [kJ/mol]	ΔS_m [J/mol K]	ΔV_m $\left[\frac{\mathrm{cm}^3}{\mathrm{mol}}\right]$	n	r_h [Å]	h^* $\left[\frac{\mathrm{mol\,H_2O}}{\mathrm{mol\,OBS}}\right]$	D_m $[10^{-10}\,\mathrm{m}^2/\mathrm{s}]$	η_el %
293	10.0	21.0	7.1	47.3	17.8	90	33	42	0.65	60
303	11.0	21.4	7.5	45.6	23.1	88	32	40	0.87	58
313	12.0	21.8	8.0	44.0	29.0	85	31	37	1.13	54
323	13.0	22.2	8.8	41.4	32.2	82	31	35	1.39	52
333	15.0	22.6	9.2	40.2	38.3	78	30	31	1.71	48

before CMC $\qquad \beta_S \times 10^{11} = 45.4237 + 0.186\varepsilon \quad \mathrm{Pa}^{-1}$, (6.27)

after CMC $\qquad \beta_S \times 10^{11} = 45.4219 + 0.169\varepsilon \quad \mathrm{Pa}^{-1}$. (6.28)

Near CMC, where the behavior of β_S is observed to have a divergence, a difference $\Delta \beta_S$ separates regular and anomalous parts:

6. Micellization as a Phase Transition

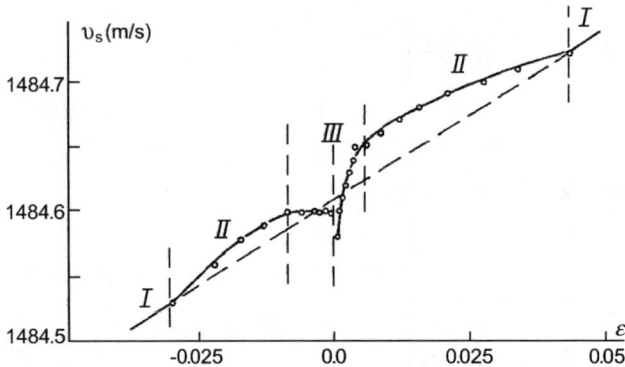

Fig. 6.6. Dependence of the sound velocity on the remoteness from CMC of the OBS–water solution at 293 K and $x_c = 9.959$ mol/m^3

$$\text{before CMC} \quad \Delta\beta_S \times 10^{14} = 1 \times 10^{-2}\varepsilon^{-1} \quad \text{Pa}^{-1}, \tag{6.29}$$

$$\text{after CMC} \quad \Delta\beta_S \times 10^{14} = 0.5 \times 10^{-2}\varepsilon^{-1} \quad \text{Pa}^{-1}. \tag{6.30}$$

Experimental values of $\Delta\beta_S$ (circles) and the values calculated as a function of concentration using (6.29) and (6.30) are shown in Fig. 6.7. It should be noted that the experimental data are significantly different from calculated ones in the vicinity of CMC, where $\varepsilon \approx 10^{-3}$.

Introduction of additional surfactant is likely to give rise to increased intermolecular correlation when saturating the solution with OBS molecules, corresponding to the beginning of a decrease in the density of the system. As a result of this process, dynamic nuclei that do not yet have a micellar structure accumulate in the system. However, the need for a certain orientation of the hydrocarbon chains conditions the formation of loosely packed structural sections. Hence, as the compressibility of the system grows, the partial molar

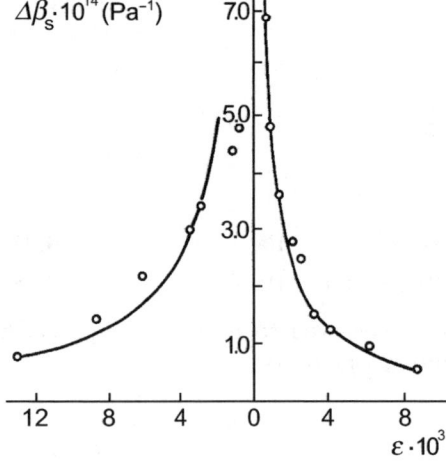

Fig. 6.7. Singular part of the adiabatic compressibility of the OBS–water solution

volume changes slightly. After reaching a certain critical size (or density), nuclei rapidly transfer into a micellar form with the formation of a unitary hydrate membrane. At the same time the compressibility and partial molar volume change abruptly. Peculiarities of this new micellar condition, such as high dispersity, finiteness of the number of particles, thermodynamic stability, semi-dispersity of sizes, and relative independence of micelles near the CMC, give rise to a continuous accumulation of micelles and influence the compressibility of solutions, as is also evidenced by the disappearance of the phase transition at a temperature of 313 K.

As has been shown, micellization begins as a phase transition of the first order when there appear nuclei of the new phase whose micro-heterogeneity interrupts its own further development. However, small specific heat and volumetric changes in independent sections throughout the solution, and proportionality of the micellar phase of the concentration over-criticality allow one to interpret this transition as a smeared phase transition of finite kind. This means that the process is not completed, and the further growth of the surfactant concentration motivates the formation of disc-like and laminate micelles, and then lamella liquid structures.

The agreement of the critical indices of density and compressibility with those predicted by the Landau theory of phase transition is not accidental. At the same time, because of the peculiarity of the phase transition under micellization conditions, the behavior of a real system diverges from the theoretical conclusions made in the vicinity of CMC.

At present, micellization is investigated in stages from the formation of dimers to large micelles. Experiments unambiguously show that in strongly diluted solutions, dimers predominate over monomers, and this region is well described by the law of mass action. Oligomers can also be described by this law, assuming that they exist in sufficient amounts. However, all these stages are investigated separately and many questions related to the kinetics of this process remain open.

6.9 Kinetics of Micellar and Pre-micellar Associations

The model presented by E.A. Anianson and C.H. Wall is of especial interest among various theoretical approaches to micellization [176]. It was developed for solutions containing surfactants above the critical micellization concentration, based on the assumption that micelles are formed from monomers via the formation of dimers, trimers and more complex aggregates. In this case, the aggregation space is divided into three regions. The first region contains monomers and dimers, while the concentrations of trimers and other oligomers are very low. The second region is related to aggregates of an intermediate dimension whose concentration is much less than the concentration of monomers and genuine micelles. It should be noted that the existence of such aggregates is not observed experimentally. The third region is that of genuine micelles.

The Anianson–Wall model allows one to obtain good qualitative agreement between two experimentally observed relaxation processes: a 'slow' process with characteristic times of 10^{-2}–10^{-4} s, due to formation and destruction of micelles, and a 'fast' process having characteristic times of 10^{-6}–10^{-8} s, conditioned by the exchange process between micelles and surfactant monomers.

The correspondence between experimental data and model predictions makes it possible to use the model in the region of concentrations less than CMC. At the present time, indirect arguments for the existence of premicellar aggregates in solutions of surfactants below CMC have been obtained by electrochemical methods, although their structure and stability are as yet uncertain.

Investigations of the acoustic spectra of sodium n-octylsulfonate and decylsulfonate solutions are described in [11, 79]. The frequency dependence of the absorption coefficient, corresponding to the square of the frequency, for solutions at various concentrations of OBS, is shown in Fig. 6.8. A decrease in the concentration of surfactant gives rise to a decrease in relaxation frequency, although the frequency increases with increasing temperature. Analysis of the experimental results revealed that these curves are described by single relaxation processes [see (4.17)]:

$$\frac{\alpha_p}{\omega^2} = \frac{A_p}{1 + (\omega/\omega_r)^2} + B_p \ .$$

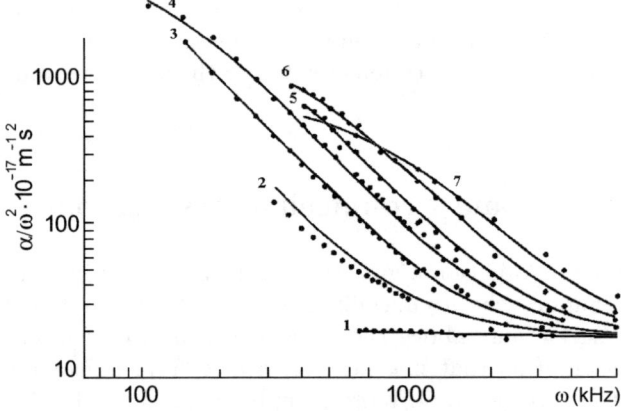

Fig. 6.8. Frequency dependence of ultrasonic absorption for various OBS concentrations at temperature 303 K: (curve 1) 0.34, (curve 2) 0.46, (curve 3) 0.65, (curve 4) 0.97, (curve 5) 1.39, (curve 6) 2.53, (curve 7) 4.91

6.9 Kinetics of Micellar and Pre-micellar Associations

Analyzing the curve at fixed temperatures and applying the equations obtained within the framework of the Anianson–Wall theory, the following relation was obtained:

$$2\pi\omega_r = \frac{k_{-1}}{\chi_d^2} + \frac{k_{-1}\tilde{x}}{\bar{n}},$$

where $\tilde{x} = x/x_1 - 1$, and x_1 and x are the monomer concentration and total concentration of surfactant molecules, respectively. k_{-1} is the dissociation rate constant, n and $\bar{n}\chi_d$ are the aggregation number and dispersion of micellar dimension, respectively. The maximal absorption due to the relaxation process is described by the relation

$$\lambda\alpha_{\text{pmax}} = \frac{A_p\tilde{x}}{T[1 + \chi_d^2\tilde{x}\bar{n}]}(1 - dx),$$

where

$$A_p = \frac{\pi\rho\vartheta_c^2(\Delta V_s)^2\chi_d}{2k_B\bar{n}}$$

is a coefficient, ΔV_s is the volume change conditioned by the exchange process, ρ is the solution density, d is a positive coefficient introduced to account for small changes in constants ΔV_s, $\bar{n}$ and χ_d occurring under changes in the surfactant concentration and allowing us to describe the revealed extremal behavior of the magnitude $\lambda\alpha_{p_{\text{max}}} = A_p\omega_r\vartheta_c/2$.

The results obtained after processing the data are shown by solid lines in Fig. 6.8. The curve showing the dependence of $\lambda\alpha_{\text{pmax}}$ on the relative concentration at 333 K is shown in Fig. 6.9. The experimental data are clearly in good agreement with theory, including the maximum in the curve $\lambda\alpha_{p_{\text{max}}}$ at $c(x - x_1)/x_1$.

The length of the hydrocarbon chain of an OBS molecule including a benzol ring is equivalent to 11.5 lengths of CH_2 groups. In this solution, acoustic relaxation was not observed at surfactant concentrations less than CMC. It could be conditioned by a low relaxation frequency (about 20 kHz near CMC) and a small CMC value. In order to carry out observations of acoustic relaxation at surfactant concentrations below CMC, we used sodium decylsulfonate with the length of a CH_2 group equal to 10 and a CMC about 4 times higher than in the previous case.

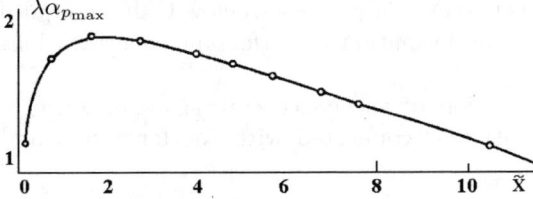

Fig. 6.9. Dependence of $\lambda\alpha_{p_{\text{max}}}$ on the OBS relative concentration

96   6. Micellization as a Phase Transition

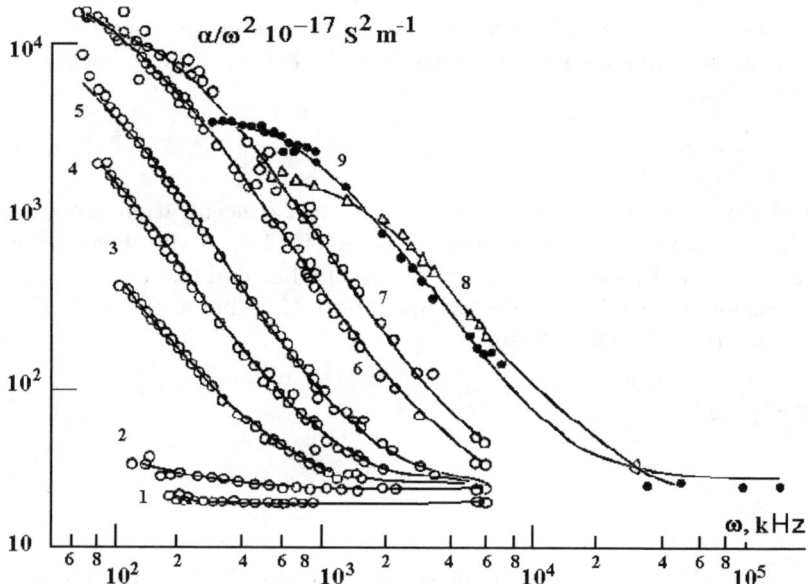

Fig. 6.10. Frequency dependence of the absorption coefficient for various surfactant concentrations at temperature 303 K: (curve 1) 0.524, (curve 2) 0.989, (curve 3) 0.963, (curve 4) 1.0, (curve 5) 1.04, (curve 6) 1.242, (curve 7) 1.49, (curve 8) 6.02, (curve 9) 4.196

Figure 6.10 shows experimental results of investigations into the frequency dependence of the absorption coefficient α_p/ω^2 for various surfactant concentrations at temperature 303 K. The acoustic technique made it possible to detect acoustic relaxation up to concentrations of 0.5 CMC. The processing of experimental data showed that all the curves (before and after CMC) can be described by single relaxation processes. Three regions occur in the concentration dependence of the acoustic relaxation frequency (Fig. 6.11). The first region, from 0.5 CMC to 0.9 CMC, is characterized by a weak monotonic increase in the relaxation frequency, the second, when the surfactant concentration approaches the CMC, by an abrupt decline, and the third, from 1 to 9 CMC, by a nearly linear monotonic increase. The third region was described within the framework of the Anianson model. The approximate results are shown in Fig. 6.11 by solid lines. To describe micellization properly, one needs to make assumptions about the character of processes below CMC, and also about their connection with micelle formation at surfactant concentrations above CMC.

By assuming that the first region of the concentration dependence of the frequency of acoustic relaxation is connected with the formation and destruction of dimers

$$A_1 + A_1 \rightleftharpoons A_2,$$

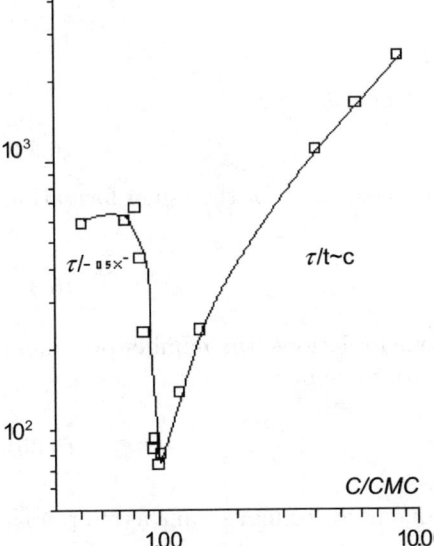

Fig. 6.11. Concentration dependence of the acoustic relaxation frequency in water solutions of OBS

where A_1 stands for surfactant monomers and A_2 for dimers. The relaxation frequency of this reaction can be found from the relation

$$\omega_r = 2\pi k^-(K + 2x_1) \,,$$

where

$$K = \frac{k^+}{k^-}$$

is an equilibrium constant, with k^+, k^- constants of the forward and reverse reactions, respectively.

The second region of the concentration dependence is marked by the formation of aggregates containing more than two surfactant molecules:

$$A_1 + A_2 \rightleftharpoons A_3 \,,$$
$$A_1 + A_3 \rightleftharpoons A_4 \,,$$
$$\vdots$$
$$A_1 + A_{n-1} \rightleftharpoons A_n \,, \quad \text{etc.}$$

To solve this system of equations, one needs to know the equilibrium distribution of concentrations between different aggregates. After solving this problem, one can completely describe the concentration dependence of the acoustic parameters.

To this end, we consider the equations characterizing an exchange involving a monomer and the nth oligomer (n-mer):

$$A_1 + A_{n-1} \rightleftharpoons A_n \,. \tag{6.31}$$

For a separate stage, the kinetic equation is written

$$\frac{dx_n}{dt} = -x_n k_n^- + x_1 x_{n-1} k_n^+ . \qquad (6.32)$$

In equilibrium the following relation must be satisfied:

$$0 = -x_n k_n^- + x_1 x_{n-1} k_n^+ . \qquad (6.33)$$

According to this definition, the equilibrium constant for the n-mer formation reaction is

$$K_n = \frac{k_n^+}{k_n^-} . \qquad (6.34)$$

By taking into account the above-mentioned relations, the number of n-mers can be expressed in terms of the number of monomers:

$$x_n = (x_1)^n \prod_{i=2}^{n} K_i . \qquad (6.35)$$

To solve the system of equations (6.31), one needs to make some assumptions about the dependence of the reaction rate constants on the degree of aggregation of n-mers. The rate constant of the forward reaction can be calculated on the assumption of a diffusion-limited forward reaction stage. According to the Smolukovski theory (6.18), this implies

$$k_n^+ = 4\pi N_A (D_1 + D_{n-1}) R^* ,$$

where D_1, D_{n-1} are coefficients of monomer and $(n-1)$-mer diffusion, respectively, and R^* is the distance at which a monomer is captured. In the first approximation, this distance can be considered to be equal to the length of the stretched monomer. By assuming that the aggregates and the monomer are close to spherical, we obtain

$$D_n = \frac{k_B T}{6\pi N_A \eta R_n} ,$$

where R_n is the size of an aggregate. Then for the constant rate of the forward reaction, the relation can be written

$$k_n^+ = \frac{2k_B T}{3\eta} \left(\frac{1}{R_1} + \frac{1}{R_{n-1}} \right) R^* . \qquad (6.36)$$

By simplifying the relation (6.36) under the assumption of small changes in molar volume during micellization, we obtain

$$V_n = n V_1 = \frac{4}{3} \pi R_n^3 .$$

This leads to

$$k_n^+ = k^+ \left[1 + (n-1)^{-1/3} \right] . \qquad (6.37)$$

6.9 Kinetics of Micellar and Pre-micellar Associations

The dependence of the reaction constant on the number of aggregations is weak. The magnitude of k_n^+ varies by a factor of about three from a dimer to the aggregate with $n = 40$. To obtain the relation for the rate constant of the reverse reaction, we assume the process of pre-micellar association to be fluctuant. The rate constant of a reverse reaction and its dependence on the number of monomers in aggregates can be obtained via the fluctuon model for micellization, according to which the stability of pre-micellar aggregates is determined by changes $\Delta\Phi$ in the thermodynamic potential [see (6.17)]:

$$\frac{\Delta\Phi}{k_B T} = -\frac{Ax}{k_B T} + \left(\gamma_1^{5/3} + f_1\right)\Theta n \approx -\frac{Ax}{k_B T} + f_1 \Theta n \ .$$

The constant describing the dissociation rate can be found from the approximate equality

$$k_n^- \approx 4\pi D_0 R_n \exp\left(-\frac{\Delta\Phi}{k_B T}\right) = 4\pi D_0 R_n \exp\frac{Ax}{k_B T}\exp(-f_1 \Theta n) \ . \quad (6.38)$$

It follows that the value of k_n^- decreases with an increasing number of monomers in an aggregate.

The value of the constant $\Theta(\sigma, t)$ depends on the surface tension of the aggregate (pre-micellar formation or micelle). Because σ tends to be small, it can be assumed that $f_1 \Theta(\sigma, t) n \ll 1$, so for k_n^- we obtain

$$k_n^- \approx \frac{k_0^-}{1 + dn} \ ,$$

where $d = f_1 \Theta(\sigma, t) n$ and $k_0^- = 4\pi D_0 R_n \exp(Ax/k_B T)$. For further analysis it is convenient to use a simplified formula reflecting the functional dependence of k_n^- on n:

$$k_n^- = \frac{k_0^-}{n} \ , \quad (6.39)$$

where k_0^- is constant. By substituting (6.37) and (6.39) into (6.34), we obtain the following relation for the equilibrium constant:

$$K_n = K\left[1 + (n-1)^{-1/3}\right] n = K F(n) \ ,$$

where $F(n)$ is a function taking into account the dependence of K on n. Then on the basis of (6.34), the concentration of n-mer is

$$x(n) = (x_1)^n K^{n-1} \prod_{i=2}^{n} F(i) \ . \quad (6.40)$$

Let us assume that a change in pressure produces a change ΔK in the equilibrium constant and that the concentrations of monomer and aggregates change from their equilibrium values to the corresponding values Δx_1 and Δx_n. From (6.40) at $n > 2$, we obtain

6. Micellization as a Phase Transition

$$\Delta x_n = x_1^n(n-1)K^{n-2}\Delta K \prod_{i=2}^{n} F(i) . \tag{6.41}$$

Changes in the concentration of monomers are calculated from the equation of material balance,

$$\Delta x_i = -\Delta K \sum_{2}^{n} nx_1^n(n-1)K^{n-2} \prod_{i=2}^{n} F(i) = -\Delta K [Z(x_1)]^{-1} . \tag{6.42}$$

The limiting values of the functions are

$$\lim_{x_1 \to 0} Z(x_1) = \frac{1}{4x_1^2} ,$$

$$\lim_{x_1 \to \infty} Z(x_1) = \left[nx_1^n(n-1)K^{n-2} \prod_{i=2}^{n} F(i) \right]^{-1} = \frac{K}{x_1(n-1)} .$$

To determine the rate of change of the n-mer concentration, (6.32) is written

$$\frac{\mathrm{d}(x_n + \Delta x_n)}{\mathrm{d}t} = -(x_n + \Delta x_n)k_n^{-1} + (x_1 + \Delta x_1)(x_{n-1} + \Delta x_{n-1})k_n^+ , \tag{6.43}$$

$$\frac{\mathrm{d}(\Delta x_n)}{\mathrm{d}t} = -\Delta x_n k_n^{-1} + (x_{n-1}\Delta x_1 + x_1 \Delta x_{n-1})k_n^+ . \tag{6.44}$$

For $n > 2$,

$$\frac{\mathrm{d}(\Delta x_n)}{\mathrm{d}t} = \left[\frac{\Delta x_n}{F(n) \cdot K} + x_{n-1}\Delta x_1 + x_1 \Delta x_{n-1} \right] k^+ \left[1 + (n-1)^{-1/3} \right] , \tag{6.45}$$

and for $n = 2$,

$$\frac{\mathrm{d}(\Delta x_2)}{\mathrm{d}t} = 2k^+ \left(\frac{\Delta x_2}{4K} + 2x_1 \Delta x_1 \right) . \tag{6.46}$$

The case $n = 1$ demands special treatment because dimerization is the first stage of the process, and also because particles A_1 take part in every reaction. The rate of change of the concentration of these particles can be found from the equation of material balance:

$$\frac{\Delta x_1}{\tau} = \sum_{i=2}^{n} i \frac{\mathrm{d}(\Delta x_i)}{\mathrm{d}t} . \tag{6.47}$$

Then the relaxation time can be calculated from

$$\frac{1}{\tau} = \frac{1}{\Delta x_1} \sum_{i=2}^{n} i \frac{\mathrm{d}(\Delta x_i)}{\mathrm{d}t} .$$

Rearranging (6.45) and (6.46) and substituting (6.41) and (6.42) for the corresponding values x_1 and x_n, we obtain

$$\frac{d(\Delta x_n)}{dt} = -\Delta K \left[x_1^n K^{n-3} \prod_{i=3}^{n-1} F(i) + x_1^{n-1} K^{n-2} \prod_{i=3}^{n-1} \frac{F(i)}{Z(x_1)} \right]$$
$$\times k^+ \left[1 + (n-1)^{-1/3} \right].$$

The rate of change of the concentration of the nth aggregate is

$$\frac{d(\Delta x_2)}{dt} = -2k^+ \Delta K \left[\frac{x_1^2}{K} + \frac{2x_1}{Z(x_1)} \right],$$

which is the rate of change of the dimer concentration.

On the basis of the above expressions, the dependence of the relaxation frequency on the aggregate association number and the kinetic parameters of micellization can be calculated as

$$\frac{1}{\tau} = \frac{1}{Z(x_1)} \sum_{i=3}^{n} x_1^{i-1} K^{i-2} \prod_{i=3}^{n-1} F(i) \Big[x_1(i-1) F(i) k^-$$
$$+ x_1^2(i-2) K k^+ + Z(x_1) k^+ \Big] + 2k^+ \left[\frac{x_1^2}{K} Z(x_1) + 2x_1 \right].$$

By considering the limits $x \to 0$ ($x \ll x_{\text{cmc}}$) and $x \to \infty$ ($x \gg x_{\text{cmc}}$), one can obtain the following relations for the dependence of the relaxation time on surfactant concentration:

- when only monomers are present, $1/\tau \sim 0.5 k^-$,
- at concentrations considerably higher than the CMC,

$$\frac{1}{\tau} \sim \sum_{i=3}^{n} x_1^{i-1} K^{i-2} \prod_{i=3}^{n-1} F(x_1^i) \sim x.$$

The above equations predict rather well the experimentally observed dependence of acoustic relaxation frequency on the concentration (Fig. 6.11). This is one more step in the description of stage-by-stage monomer aggregation from dimers to the formation of micellar structures, and elucidation of their dynamical changes, within the framework of a unified model.

6.10 Micellization Under Intensification of Molecular Mass Transfer

The presence of biphilic components of low molecular weight in a solution influences the structure of water and also the self-assembly process for micellar structures [179]. The peculiarity of evolutive concentration fluctuations of amphiphilic (surfactant) and biphilic (medium-chain alcohol) molecules determines the dynamical characteristics of the process up to the appearance of micelles. At the same time, differences in geometrical dimensions of molecules and corresponding coefficients of diffusion of the above-mentioned components, as well as the low density of solute particles, can lead to simplifications such as the space and time independence of concentration fluctuations.

Solubilization of polar substances in water is accompanied by the appearance of a new molecular order and characterized by certain types of interactions. In dilute solutions of alcohols, changes in solubilization can be so significant that one may observe an order–disorder phase transition even in the absence of a phase-separation point. The experimental facts can be satisfactorily explained if one takes into account the growth of a near-order scale with simultaneous ordering of alcohol molecules in a quasi-crystalline structure of water. The latter seems to be a manifestation of the peculiar character of the intermolecular interactions of solute particles.

Measurements of density ρ, dynamical viscosity η and ultrasound velocity ϑ_c for aqueous solutions of n-butanol at frequency 2 MHz in the temperature range 293–333 K were carried out in order to study changes in near-molecular order with variation of surfactant concentration in the range of 0–0.8 mol/dm^3. Some parameters of the investigated system were calculated using known thermodynamic relations, to facilitate analysis of the obtained results (Table 6.4).

Table 6.4. Thermodynamic parameters for aqueous solutions of n-butanol

T [K]	$d\rho/dx$ $\left[\dfrac{\text{kg/m}^3}{\text{mol/dm}^3}\right]$	V_μ^∞ $\left[\dfrac{\text{cm}^3}{\text{mol}}\right]$	ΔV_μ^∞ $\left[\dfrac{\text{cm}^3}{\text{mol}}\right]$	R_B [Å]	α_V^∞ [10^{-4}/K]	α_V^0 [10^{-4}/K]	$d\eta/dx$ $\left[\dfrac{10^{-3}\text{Pa·s}}{\text{mol/dm}^3}\right]$
293	−11.375	85.65	−5.8	2.59	8.0	8.0	0.295
303	−11.875	86.35	−5.8	2.60	8.5	9.0	0.220
313	−12.375	87.15	−5.9	2.61	9.5	9.5	0.125
323	−12.875	88.15	−5.9	2.62	7.5	10.0	0.115
333	−12.500	88.30	−6.7	2.63	2.5	11.0	0.080

The solution density at temperatures 293–323 K was found to have a linear dependence in the range of concentrations 0–0.3 mol/dm^3, and in the whole range of concentrations at the temperature 333 K. The values of $d\rho/dx$ and the partial molar volume of butanol V_μ^∞ at its limit dilution, change in the molar volume of butanol ΔV_μ^∞ due to transfer into a completely dilute solution, van der Waals radius of butanol molecules R_B, volumetric extension coefficient α_V and others, were obtained by taking into account the linearity of ρ in x.

The absolute value of $d\rho/dx$ increases monotonically to 323 K and has an abrupt change at 333 K, whilst other calculated parameters also change in an anomalous way. This is due to the fact that, in the vicinity of the temperature 333 K, the structure of the solution is significantly changed because the solubility of butanol decreases as the temperature increases up to 333 K, whereas after 333 K the solubility increases, i.e., the enthalpy of mixing becomes different too. It can be assumed that for 293–333 K (up to 0.3 mol/dm^3), the added butanol is built into the quasi-crystalline structure of water and directly takes part in the formation of structural cages by means of an OH group, with the hydrocarbon tail in cavities, i.e., a hydrophobic interaction occurs and the value of α_V^∞ is practically the same as the value of α_V^0 in pure butanol. At 333 K, butanol becomes a hydrophilic addition in the case of short-chain alcohols because of a thermal smearing of the water structure, and the value of α_V^∞ is almost the same as for the environment. Simultaneously, a sharp change $\Delta V_\mu^\infty d\eta/dx$ is observed. This analysis is also confirmed by the linear dependence of the viscosity on the butanol concentration at 333 K, as in ideally mixing systems, and by the relatively small value of $d\eta/dx$ at the same temperature, showing an insignificant change in the activation energy of viscous flow, i.e., an approximate equivalence of the relative locations of molecules. Thus, when passing from the temperature of solubility inversion (333 K) to higher temperatures, the contribution made by the hydrocarbon part of butanol molecules to the energy of intermolecular interaction and to the level of organization of molecules in the solution is compensated by an increase in the intensity of thermal motion of particles and the structure is smeared.

The structure of solutions also gets noticeably broken at butanol concentrations above 0.3 mol/dm^3. In accordance with experimental results, a faster than linear growth of the density and viscosity is observed, beginning at the concentration 0.3 mol/dm^3 (except for 333 K). The adiabatic compressibility β_S of solutions decreases ($T = 239$ K); see Table 6.5.

Table 6.5. Concentration dependence of adiabatic compressibility of aqueous solution of n-butanol

X [mol/dm^3]	β_S [10^{-12} Pa^{-1}]
0	45.6
0.05	45.45
0.1	45.30
0.2	45.00
0.3	44.65
0.5	43.85
0.8	43.05

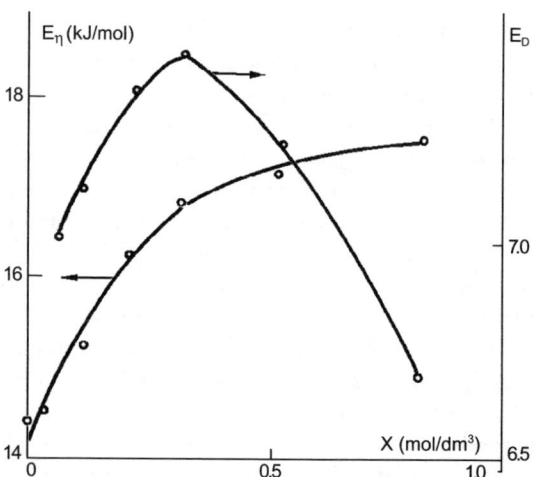

Fig. 6.12. Concentration dependence of the activation energy of viscous flow processes and dielectric polarization in aqueous butanol solution

The concentration dependence of the activation energy of viscous flow E_η and dielectric polarization E_D calculated with the Debye equation are shown in Fig. 6.12. Analysis has revealed that the increase in E at butanol concentrations above 0.3 mol/dm^3 is due to an increase in the density of molecular packing and not to the appearance of specific interactions. At the same time, the increase in E_D points to the fact that the near-order around molecules gets broken.

For the water–OBS–n-butanol system, the sound propagation velocity was measured with accuracy up to 1 cm/s by an interferometer designed jointly with researchers at the Kaunas Polytechnic Institute. Solution densities were measured by the hydrostatic weighing method with an accuracy of 5×10^{-5} g/cm^3. These solution parameters were investigated in the range of OBS concentrations 0–16 mmol/l with additions of butyl alcohol 50–800 mmol/l at temperatures 293–333 K. It should be noted that the choice of the upper solution concentration was limited by the solubility of the water–butanol mixture at room temperature.

One of the most structurally sensitive parameters is known to be the adiabatic compressibility $\beta_S = 1/\rho\vartheta_c^2$. To process jointly data on density and sound velocity, we use the dimensionless concentration ε, where $\varepsilon = (c_{\text{cmc}} - c)/c_{\text{cmc}}$. Curves showing the adiabatic compressibility dependence on the OBS concentration (0.05 mol/dm^3 of butanol), using the density approximation, are given in Fig. 6.13. The calculated value of the critical exponent is $\gamma = 1.08 \pm 0.11$, in good agreement with the value obtained from the Landau theory of phase transitions.

Note that the regions of change in adiabatic compressibility when approaching CMC for solutions containing alcohol are wider than for solutions

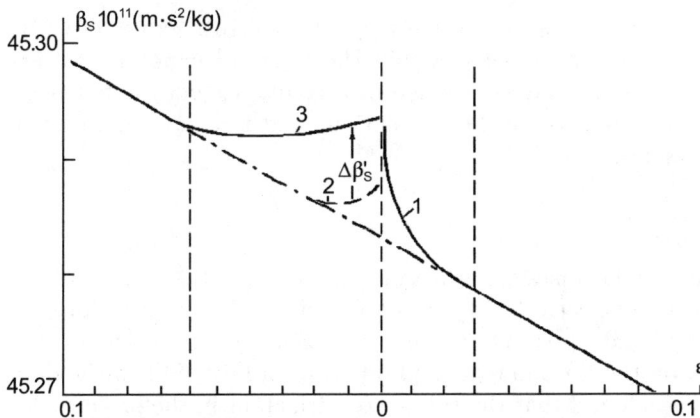

Fig. 6.13. Adiabatic compressibility dependence of the OBS–water–butanol solution. Curves 1 and 3 are experimental, and curve 2 is calculated

without alcohols [180]. This may be due not only to the presence of a wide fluctuation region for alcohol–water solutions, but also to a greater number of variants on micelle packings in the presence of a medium-chain alcohol, when possible deviations in micelle sizes from their average values (dispersion of the association number) increase considerably.

6.11 Micellization in the Electric Field of Charged Admixtures

The contribution to the electric component of free energy of micellization in solutions of ionic surfactants is mainly determined by the Coulomb interaction. Aggregation of charged amphiphilic particles of the same sign under hydrophobic interaction is limited by the efficiency of the influence of Coulomb forces. This determines the size of the existence region for stable micelle formation. Together with the developed interface of micelles having the structure of a double electric layer, the above-mentioned facts attest to the important role of electric forces in the systems under investigation.

We studied an aqueous solution of octylbenzol sodium sulfonate (OBS) with various additions of NaCl, in particular rheological, electric, superficial and acoustic properties at temperatures 293–343 K. The shear viscosity coefficient was measured by a capillary viscosimeter, density by the method of hydrostatic weighing at small surfactant concentrations in the vicinity of CMC, and by the picnometric method at large concentrations, conductivity by the compensation method, and surface tension coefficient by differential depression of bubbles.

The curves describing the dependence of the studied properties on surfactant concentration were shown to have a characteristic break corresponding to the critical micellization concentration (Figs. 6.14–6.16). Values of CMC found by different methods coincide within the range of experimental accuracy. Assuming that the partial volume of a solute V_n does not depend on the concentration, at least in dilute solutions and in a narrow range of concentrations, we obtain

$$V_n = \frac{M(1 - \partial \rho / \partial x)}{\rho_0},$$

where M is the molecular mass of surfactant, ρ_0 and ρ are the density of the solvent and solution, respectively, and x is the surfactant concentration.

Using the calculated apparent volumes of OBS, one can calculate the influence of NaCl on the thermodynamic properties of OBS. When the data was processed, it was found that the volume of surfactant in the pre-micellar phase goes through a minimum when a salt is added (Table 6.6).

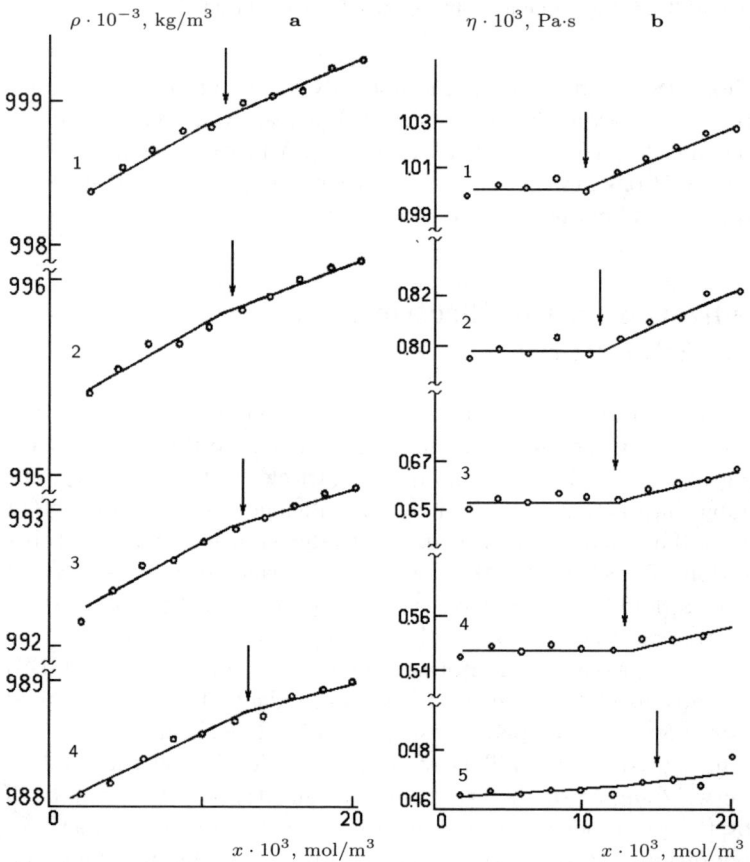

Fig. 6.14. Dependence of density (**a**) and shear viscosity (**b**) of the OBS–water solution at various temperatures: (curve 1) 293 K, (curve 2) 303 K, (curve 3) 313 K, (curve 4) 323 K, (curve 5) 333 K

6.11 Micellization in the Electric Field of Charged Admixtures

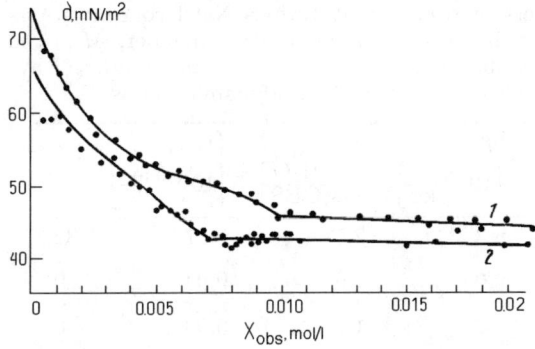

Fig. 6.15. Dependence of surface tension on OBS concentration in OBS–water + NaCl at 303 K and NaCl concentrations 0.005 mol/l (1), 0.015 mol/l (2)

From the behavior of viscosity above CMC one can determine a 'hydrodynamic' volume of the micellar phase per mole of OBS in the micellar state, within the framework of the Onsley hydrated particle theory:

$$G_v = [\eta] n B N_A M_{OBS},$$

where B is a constant depending on the form of the rigid colloidal particles, n is the association number, M_{OBS} is the molecular mass of OBS, N_A is the Avogadro constant, and $[\eta]$ is the characteristic viscosity of the solution above CMC, calculated using the formula

$$[\eta] = \lim_{x \to x_c} \frac{1}{x - x_c} \ln \frac{\eta}{\eta_c}.$$

Regarding the micelles as spherical particles, the value of B was taken as 2.5.

When a small amount of salt is added, the 'hydrodynamic' volume of a micellar surfactant is seen to decrease abruptly. This is likely to be due to

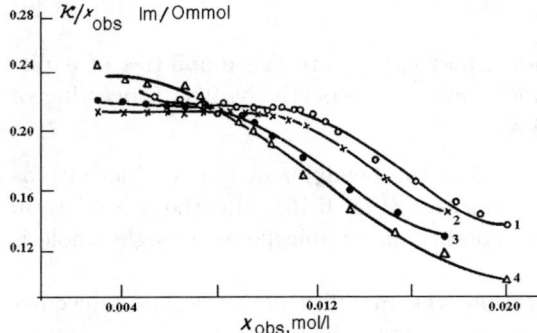

Fig. 6.16. Dependence of conductivity on OBS concentration in OBS–water + NaCl at various NaCl concentrations [mol/l]: (curve 1) 0.0, (curve 2) 0.002, (curve 3) 0.01, (curve 4) 0.02

6. Micellization as a Phase Transition

Table 6.6. Rheological parameters of solutions at various NaCl concentrations x_{NaCl}. G_V is the 'hydrodynamic' volume, n is the association number, M is the molecular mass of associates, $[\eta]$ is the intrinsic viscosity, δ is the number of hydratations, D_m is the diffusion coefficient, r_h is the hydrodynamic radius

x_{NaCl} $\left[\frac{\text{mol}}{\text{dm}^3}\right]$	G_V $[10^{-4}\,\text{Å}^3]$	n	M $\left[\frac{\text{kg}}{\text{mol}}\right]$	$[\eta]$ $\left[10^{-3}\frac{\text{m}^3}{\text{kg}}\right]$	δ $\left[\frac{\text{kg}\,\text{H}_2\text{O}}{\text{kg}\,\text{OBS}}\right]$	D_m $\left[10^{-10}\frac{\text{m}^2}{\text{s}}\right]$	r_h [Å]
0	13.72	88	25.73	8.3	2.5	0.87	32
5	10.72	92	26.90	6.0	1.6	0.94	29
10	10.95	91	26.90	6.2	1.7	0.90	30
15	8.60	89	26.00	5.0	1.2	1.02	27
20	8.35	86	25.10	5.0	1.1	1.03	27

a decrease in the hydrate shell of micelles on account of a redistribution of water between micelles and salt ions. With a further increase in the amount of NaCl, it goes down slowly owing to a decrease in the number of associations.

The data on the conductivity of salt solutions of surfactants were processed assuming independence of the contribution of dissociation products from surfactants and NaCl to the conductivity. After micellization, the micellar phase mobility as a charge carrier was also taken into account. Note that the contribution of the ionic 'fur' surrounding a micelle was not considered in the first approach. The following relation can be written for equivalent conductivity:

before CMC $\quad k/x = (k_{\text{OBS}} + k_{\text{NaCl}})/x$,

after CMC $\quad k/x = (k_{\text{sur}} + k_{\text{NaCl}} + k_{\text{mic}})/x$.

The last equation can be transformed to

$$\frac{k}{x} = \frac{v_p x_c}{x} + \frac{e v_m (x - x_c)}{x} , \qquad (6.48)$$

where e is the charge on the ions, and v_p, v_m are the mobilities of a dissociated radical of surfactant and a micelle, respectively. The processing of experimental data has shown that:

- equation (6.48) adequately describes the behavior of the conductivity as a function of the amount of surfactant (Fig. 6.16), and the second term of the equation related to the mobility of the micelle as a single whole is significant;
- with increasing salt concentration, the mobility of surfactant molecules remains constant, within the range of experimental error;
- the mobility of micelles decreases significantly with increase in the salt concentration.

6.11 Micellization in the Electric Field of Charged Admixtures

It has been shown that, after adding a salt, the 'hydrodynamic' volume of the micellar phase decreases. This is due to a decrease in the mobility of micelles. Apparently, this can be explained by the influence of the relaxation effect of a diffusion layer consisting of counter-ions [182]. The values of the mobility of the surfactant molecules and micelles depending on the amount of NaCl are given in Table 6.7.

Table 6.7. Mobility of monomers v_1 and micelles v_m at different salt concentrations

X_{NaCl} [m.f.]	mobility	
	$v_1 \times 10^{-6}$	$v_m \times 10^{-7}$
0	2.0	6.1
0.002	2.3	7.5
0.010	2.4	7.0
0.015	2.5	5.8
0.020	2.5	4.0

The experimentally constructed phase diagram is shown in Fig. 6.17. It should be noted that an abrupt rise in the Kraft point and a slow lowering of the CMC are observed when adding the salt. Both these factors determine the boundaries of the region of micellar structures via temperature and concentration. As a result, the region of molecular solubility of OBS in the phase diagram considerably decreases.

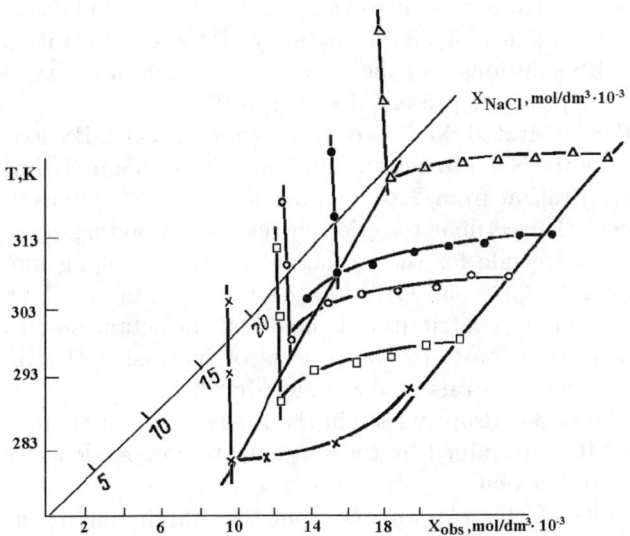

Fig. 6.17. Phase diagram of the OBS–water + NaCl system

Measurements of the sound velocity at frequency 14.85415 MHz using the interferometric method have shown that it goes through a minimum at CMC (Fig. 6.18). The character of the change in the sound velocity is related to the peculiar behavior of both the solution density and the adiabatic compressibility β_S near CMC. With addition of a salt, the compressibility of the solution decreases and the CMC does not shift, as observed when adding alcohols, but the region of phase transition tends to widen.

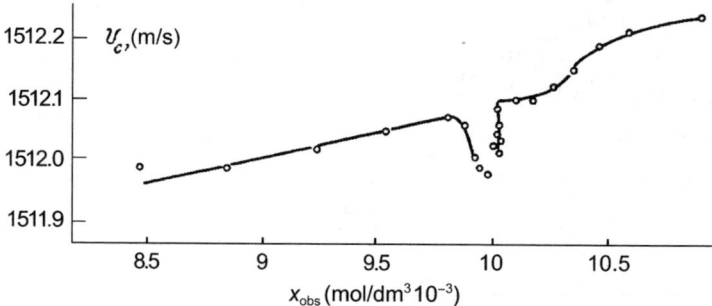

Fig. 6.18. Dependence of sound velocity on OBS concentration in the OBS–water + NaCl solution

The above results focused primarily on the process of solute particle association. At the same time, a change in the structure of water is not the least peculiarity observed amongst the physical properties of the solutions investigated. This change is evidenced by the structural luminescence method (SL), whose spectral properties are caused by distortions in the structure of disordered media [181]. The dependence of the SL intensity on the concentration of aqueous NaCL and OBS solutions and the curve showing the intensity of the combination scattering peak are presented in Fig. 6.19.

When dissolving OBS, hydrated Na^+ ions and asymmetrical OBS ions emerge in water. Increasing the NaCl concentration from 1.7 to 8.6 mmol/dm^3 and the surfactant concentration from 2 to 8 mmol/dm^3, the SL intensity increases, and the slope of the rectilinear section in the corresponding curve is larger by an order of magnitude for the surfactant solution. Taking into account the influence of Na^+, Cl^-, and OBS ions on the structure of H_2O, one can conclude that a primary contribution is made by surfactant anions, in particular their hydrocarbon part, to the process of increasing the SL intensity when the concentration is raised to 8 mmol/dm^3.

Further in curve 2, the slope abruptly rises in the range 8–10 mmol/dm^3, which is the region of CMC determined by the concentration dependence of the viscosity and ultrasound velocity.

Let us view the results of this experiment from the standpoint of the fluctuon model. When dissolving NaCl in water, the active emergence of Na^+ ions has no impact on the luminescence. By dissolving OBS, as in the

6.11 Micellization in the Electric Field of Charged Admixtures

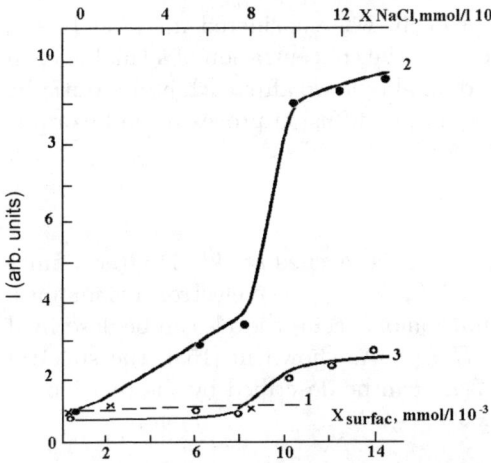

Fig. 6.19. Dependence of the SL intensity on the concentration of aqueous NaCl (1) and OBS (2) solutions at wavelength 540 nm, and dependence of the intensity of the combination scattering peak (CS) of deformation oscillations of H_2O molecules with wave number $1\,640$ cm^{-1} on the concentration of OBS solutions (3)

first case, Na$^+$ ions split off, but the electron retained on the benzol ring of an OBS molecule is, in fact, delocalized over the whole surfactant molecule. As shown above, during the conformation of a hydrocarbon part of an OBS molecule, a local state arises in contact with the water, and hence a weakly bound electron is retained in the water after dissociation of Na$^+$ ions. In their turn, Na$^+$ ions solubilized in water create favorable conditions for the appearance of small acceptor levels (Fig. 6.20).

Hence, the end result of solubilization of an OBS molecule in water is the formation of distributed donor–acceptor couples at a distance of R from each other. A surfactant molecule with a localized electron in a local level corresponds to a donor center, whilst acceptor centers are formed by water molecules and Na ions. Hence, $\tau(R) = \tau_0 \exp(2R/a)$, where $\tau \approx 10^{-8} a$ s,

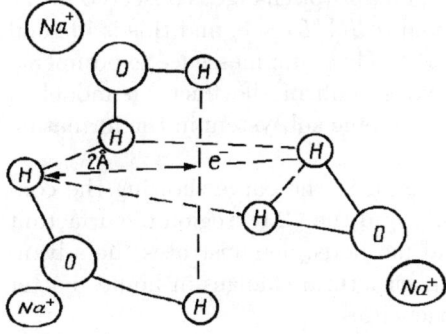

Fig. 6.20. Model of a structural trap for an electron created by three molecules of H_2O and three Na$^+$ ions

and a is the Bohr radius. Such a problem was considered in [183]. It was shown that the wandering electron caused the concentration of available local acceptor centers to decrease by a factor of about two after each jump. Initially, the available states are close. Consequently, diffusion proves to be favorable and the following condition is observed:

$$R, r \ll R_c = \frac{a}{2} \ln(r_0 \tau_0) \,.$$

However, the opposite condition $R, r \gg R_c$ is reached at $(k+1)$ after a finite number of jumps. At the same time, if $R_k < r_{k+1}$, the electron recombines. Let us assume that the energy distribution of acceptor levels can be described by the expression $g(E) = g_0 \exp(-E/E_0)$. As shown in [158], the spectral density of stationary luminescence $I(E)$ can be described by the relation

$$I(E) \sim \frac{1}{E_0} \exp \frac{0.4(E_c - E)}{E_0} \,,$$

with $E \gg E_0$, where

$$E_c = E_0 \ln \left(\frac{4\pi}{3} R_c^3 x_a \right) \,,$$

and x_a is the concentration of acceptor levels. However, this relation needs to be supplemented by a condition that takes into account the ionization probability of a donor level of an OBS molecule with a conformation change. The following relation holds for the spectral intensity of luminescence:

$$J(E) \approx J_0 s_i \frac{4\pi x}{3 E_0 V} L^3 \exp \frac{0.4(E_c - E)}{E_0} \,,$$

where s_i is the ionization cross-section, L is the size of an OBS molecule, J_0 is the intensity of laser light, and X is the concentration of OBS molecules.

Therefore, $J(E)$ proves to be a linear function of the OBS concentration that is actually reflected in the experimental dependence. In the case of micellization, the ionization cross-section s_i for a micelle is less ($s_i^m < s_i$) due to the stronger binding of the electron with the donor center (the depth of local levels of a micelle is larger), but the size of a micelle $R_0 \sim 2L$.

As a result, at the micellization point, an abrupt change is observed in the luminescence intensity that is proportional to $R_0^3/L_3 \sim 8$, and this is in good agreement with experiment (see Fig. 6.19). Thus, luminescence experiments on water solutions of surfactant molecules confirm the fluctuon model of micellization and the active role of the electronic subsystem in the formation of micelles as a whole.

The correlation between the break region in the curve showing the concentration dependence of the SL intensity and the CMC region of surfactant solutions, established by thermophysical methods, demonstrates the advantages of SL spectroscopy when studying structural changes in liquid matter and when diagnosing dynamical heterogeneities.

7. Fluctuation Mechanism of Forced Spinodal Decomposition

7.1 Spinodal Decomposition as a Model for Microemulsion Formation

Investigations into physical properties and mechanisms of microemulsion formation require us to address the common problem of phase transformations and structurization of highly non-equilibrium systems. It is of key importance to explain the similarity of self-organization processes and stability of regular structures with phase transitions.

It can be confidently asserted that microemulsions arise as thermodynamically stable structurized states through consecutive decompositions and transformations of metastable states. One of the major properties of metastable states is their finite lifetime as a result of decomposition via fluctuation formation and growth of nuclei of competing phases. Hence, structurization of microemulsions and their stability are tightly connected with the nature of fluctuation processes.

Another peculiarity of microemulsions is their proximity to a critical state. In this connection, order parameter fluctuations must be correlated at large distances, which contradicts the demands of structurization within the system. Apparently, this is an excitation relaxation specific to such states. To understand it, one needs to take into account the correlative nature of the order parameter fluctuation, which seeks to make a system maximally homogeneous, and the ongoing non-linear process of emergence and growth of microheterogeneous fluctuations, which causes the system as a whole to divide into parts.

In statistical physics, there is no adequate theory of the dynamics of metastable states in non-equilibrium. Here the difficulty is due to the fact that one has to apply methods of equilibrium statistical thermodynamics to non-equilibrium phenomena, i.e., phenomena occurring outside the region of thermodynamic stability. In these conditions, simulation methods play a special role. It is important to make a proper choice of model that most completely reproduces the investigated object. As such a model, one can take the interface of two liquids if a mixed state in its vicinity is under conditions close to spinodal decomposition.

In the vicinity of the spinodal, the response functions of a metastable system, viz., isothermal compressibility, isobaric heat capacity and others, all increase, i.e., the behavior of thermodynamic values is the same as for critical phenomena. Moreover, a system state near the spinodal is adequately described by pseudo-critical indices [184] despite the fact that the system itself is far from a critical point. Such an analogy is possible as the susceptibility of a system at spinodal points tends to infinity and, correspondingly, the volumetric integral of the pair correlation function of the density diverges. This means that fluctuations become long range and the system effectively approaches pseudo-critical states. It should be noted that it is not yet understood how far the analogy between critical phenomena and those near the spinodal can be taken. Obviously, the question about the size of the spinodal region, where real penetration of the external perturbation takes place, must be correlated with order parameter fluctuations and with the peculiarities of metastable states.

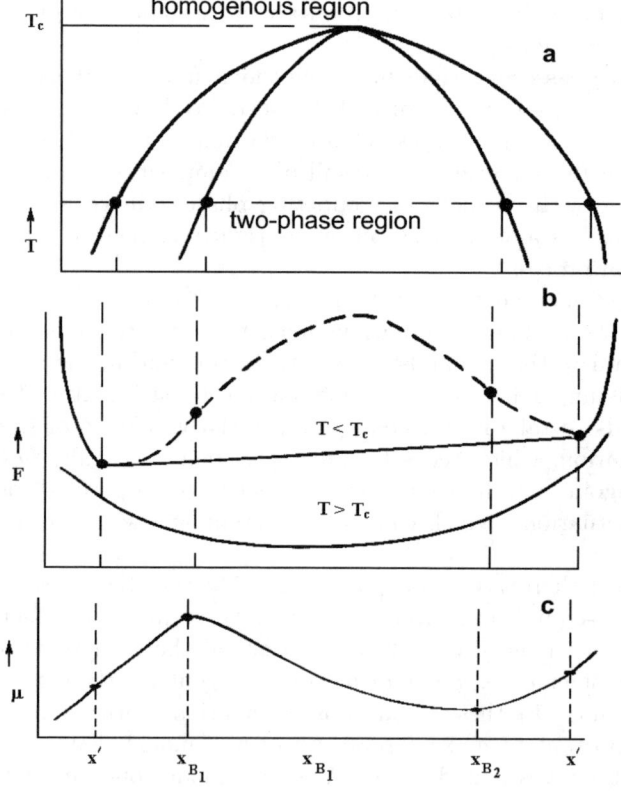

Fig. 7.1. (a) Phase diagram in a binary mixture. Dependence of free energy (b) and chemical potential (c) on composition

7.1 Spinodal Decomposition as a Model for Microemulsion Formation   115

As shown in [46, 106, 185], the magnitudes of energy barriers of formation for a new phase, which in fact determine the depth of a possible penetration by an external field into the region of metastable states, are determined by the Ginzburg number, taking into account non-locality of the fluctuation interaction. In systems with a small value of the Ginzburg number, the penetration of external forces into the region of absolute instability can occur under real experimental conditions. In this case, the growth of a new phase near the spinodal will materialize through the formation of structures. This process develops as a result of complex dynamical interaction of a) correlated composition fluctuations in a system and b) the homeophase fluctuations of the order parameter. The latter are related to the closeness of metastable states to the boundary of stability. Thus, the behavior of a system near the spinodal for the corresponding choice of object can simulate certain properties of microemulsions and the mechanisms that give rise to them. Mixtures of weakly solubilized liquids reproducing the oil–water interface can constitute such objects.

Let us consider the behavior of such a system. Under the conditions of thermodynamic equilibrium, a mixed state with breadth of order the correlation length is established on the interface of weakly solubilized liquids. Suppose the system has been transformed into a non-equilibrium state as a result of an external perturbation of thermodynamic parameters. Then the size of a mixed state on the interface can turn out to be much more than the correlation length. In this case, if frozen at the time $t = 0$, an initially disordered system can separate from the one-phase state into ordered regions of macroscopic size. The equilibrium condition demands coexistence of these regions of both phases with concentrations x', x'' (Fig. 7.1a). In the temperature–concentration space these stable concentrations form a so-called coexistence curve (Fig. 7.2).

The maximal temperature T_c of the heterogeneous state stability is defined as the critical temperature, and corresponding to it, the stable concen-

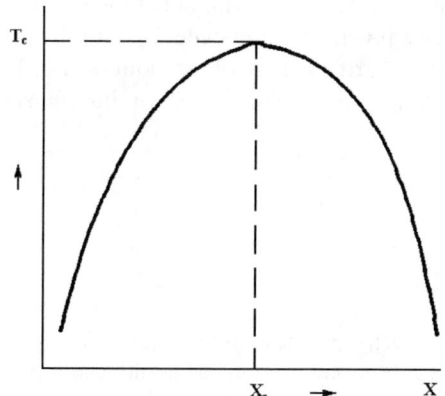

Fig. 7.2. Coexistence curve of a binary mixture

tration x_c is the critical concentration. The free energy (Fig. 7.1b) decreases with an increase in the concentration x_B in the one-phase region (entropy of mixing) and depends linearly on x_B (the amount of two coexisting phases changes) in the two-phase region.

The simplest theory of the kinetics of these processes begins by introducing the free energy F of one-phase states in a two-phase region [185]. The regime where $F > F'$ but $(\partial^2 F/\partial x_B^2)_T > 0$ still holds is called 'metastable', whereas the regime where $(\partial^2 F/\partial x_B^2)_T < 0$ is called 'unstable'. The locations of inflection points $(\partial^2 F/\partial x_B^2)_T = 0$ form a 'spinodal curve'. Taking this into account, various transition mechanisms are distinguished by phase stability with respect to continuous changes in the state at small perturbations of concentration and energy in macroscopic regions of the system. The state is considered to be stable if the phase has a regenerating reaction such that the perturbation disperses. As regards unstable states, by definition, the growth of small perturbations occurs as a result of the system response to them. Metastable states are supposed to be stable with respect to continuous changes in the state, and to lie adjacent to the region of absolute instability when solutions are oversaturated.

Figure 7.3 shows the diagram of a one-component liquid–vapor system in coordinates of specific volume $v = 1/\rho$ and temperature. The binodal curve is determined by identity of chemical potentials in coexisting phases $\mu'(P, T) = \mu''(P, T)$.

The spinodal divides regions of positive and negative values of derivatives $(\partial P/\partial \rho)_T$, $(\partial T/\partial S)_P$ tending to zero on the spinodal. This implies divergence of the thermodynamic response functions, viz., isothermal compressibility $(\partial P/\partial \rho)_T$, $(\partial T/\partial S)_P$, and isobaric heat capacity C_P. The condition $(\partial P/\partial \rho)_T = 0$ also shows that $(\partial \mu/\partial \rho)_T = 0$ or $(\partial \mu/\partial \rho)_T = 0$, since for one-component systems

$$\left(\frac{\partial P}{\partial \rho}\right)_T = \rho \left(\frac{\partial \mu}{\partial \rho}\right)_T,$$

where $\rho = Mn$, and M is the mass of a molecule. As the substance is in a labile state $(\partial P/\partial \rho)_T < 0$, it rapidly loses its spatial homogeneity and, by relaxing, gains a peculiar grain-cellular structure without phase boundaries. In a later stage, ordinary heterogeneity arises when one of the coexisting phases

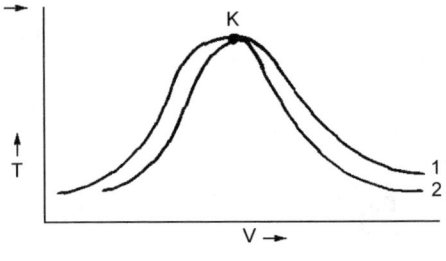

Fig. 7.3. Binodal (1) and spinodal (2) in a one-component liquid–vapor system. K is the critical point

7.1 Spinodal Decomposition as a Model for Microemulsion Formation

is dispersed in the presence of a stable surface layer. Spinodal decomposition is viewed as the development of a continuous heterogeneous structure as a result of thermodynamic instability in the system. The kinetics of such a process in a one-component system was studied for the first time in [186].

In two-component systems, the breaking of stability is related to local deviations from the equilibrium concentration. If randomly emerging heterogeneities in the composition do not disperse, but instead are amplified by the response of the system, then 'ascending' diffusion ($D < 0$) occurs there. In a two-component solution, diffusion instability precedes mechanical instability [109]. The condition of diffusion stability is written as ($T, P =$ Const.):

$$\frac{\partial \mu}{\partial x_1} > 0,$$

where x_1 is the mole fraction of the first component. In Fig. 7.1c, the condition of diffusion stability is seen to be satisfied for concentrations $x_B < x_{B1}$ or $x_B > x_{B2}$.

In Fig. 7.1b, the dependence of F on the concentration is plotted schematically for two temperatures (below and above the critical temperature), under the assumption of continuity and analyticity of the function F. The lower curve ($T > T_c$) is everywhere convex toward the axis of compositions. This provides the diffusion stability of the solution over the whole range of concentrations, which makes separation of the system into coexisting phases impossible. The upper curve has convex and concave sections. The points of inflection relate to the stability boundary of homogeneous states.

Let us consider the phenomenological theory of spinodal decomposition developed by J.W. Cahn and J.E. Hillard [187, 188] and restated in [185]. The free energy of a two-component isotropic solution can be written as

$$F = \int \left[f(x) + K(\nabla x)^2 \right] dV, \tag{7.1}$$

where $f(x)$ is the free energy density of a homogeneous solution with the composition x, $K(\nabla x)^2$ is the first non-vanishing term in the Taylor expansion of $f(x(r))$ with respect to r, describing the contribution of spatial correlation effects to the free energy.

Limiting the expansion to the second order term is equivalent to assuming that the radius of action of intermolecular potentials is much less than characteristic lengths at which the component concentration changes significantly. The molar volume is assumed to be independent of the system composition and $K > 0$ if a homogeneous state is stable above spinodal. Globally, the value of K may depend on the concentration.

If the composition at each point of the solution is not much different from the average x_0, the expansion of $f(x)$ in powers of $x - x_0$ can be limited to the quadratic term. By taking into account the condition $\int (x - x_0) dV = 0$, we find for the difference between the free energy in the solution with concentration fluctuations and that in a quite homogeneous solution with composition x_0,

$$\Delta F = F - f(x_0)V = \int \left[\frac{1}{2}\frac{\partial^2 f}{\partial x_0^2}(x-x_0)^2 + K(\nabla x)^2 \right] dV .$$

To analyze the stability of a solution with respect to infinitesimally small changes in the composition, it is convenient to employ the Fourier representation. Having expanded $x - x_0$ in the Fourier series

$$x - x_0 = \sum_k A(k) \exp(ikr) ,$$

we obtain

$$\Delta F = \frac{1}{2}V \sum_k |A(k,t)|^2 \left(\frac{\partial^2 f}{\partial x_0^2} + 2Kk^2 \right) .$$

The solution is stable with respect to infinitesimally small changes in the composition if, for all values of k,

$$\frac{\partial^2 f}{\partial x_0^2} + 2Kk^2 > 0 .$$

Since $K > 0$, this condition is always satisfied for combinations with $\partial^2 f/\partial x_0^2 > 0$ (the temperature exceeds the spinodal temperature). Inside the spinodal, i.e., in the region where $\partial^2 f/\partial x_0^2 < 0$, the magnitude of $\partial^2 f/\partial x_0^2 + 2Kk^2$ takes negative values if $k < k_c$, where

$$k_c = \left[-\frac{1}{2K}\left(\frac{\partial^2 f}{\partial x_0^2}\right) \right]^{1/2} .$$

Thus, inside the spinodal the solution is unstable with respect to infinitesimally small composition fluctuations with wavelengths $\lambda > \lambda_c = 2\pi/k_c$. The kinetics of diffusion decomposition can be determined from the solution of the diffusion equation. The difference between the chemical potentials is expressed via the functional derivative $\delta F\{x(r)\}/\delta x(r)$ as follows:

$$\mu_1 - \mu_2 = \frac{1}{V}\frac{\delta F}{\delta x_1} ,$$

where the previously introduced function F is used to calculate $\delta F/\delta x_1$. This leads to

$$\boldsymbol{J}_1 = -L\nabla \left[\frac{\partial f}{\partial x_1} - 2K\nabla^2 x_1 - \frac{\partial K}{\partial x_1}(\nabla x_1)^2 \right] . \tag{7.2}$$

Let us calculate the divergence of both sides of (7.1), neglecting the dependence of K on the composition and making use of the continuity equation. Then,

$$\frac{\partial x}{\partial t} = L\nabla^2 \left[\frac{\partial f}{\partial x} - 2K\nabla^2 x \right] . \tag{7.3}$$

In the early stages of spinodal decomposition, the composition fluctuations are small. Therefore only terms linear in x need be kept on the right hand side of (7.2) and we arrive at the linearized diffusion equation

$$\frac{\partial x}{\partial t} = L\left[\frac{\partial^2 f}{\partial x_0^2}\nabla^2 x - 2K\nabla^4 x\right],$$

whose solution is taken as a Fourier series

$$x(r,t) - x_0 = \sum_k A(k,t)\exp(ikr),$$

where $A(k,t) = A(k,0)\exp\bigl(R(k)t\bigr)$ and

$$R(k) = -Lk^2\left[\frac{\partial^2 f}{\partial x_0^2} + 2Kk^2\right],$$

is a gain factor.

The linearized theory of spinodal decomposition based on the linearized diffusion equation describes phase separation very approximately only in the early stages, when the deviation of the composition from its average value is small. It is clear that an adequate description of spinodal decomposition must be based on the linearized equation with fluctuations, whose solutions would describe a gradual transition from the phase-separation mechanism to nucleus formation. Nevertheless, to simulate the formation of microheterogeneous states such as microemulsions, the Cahn–Hillard theory may still be useful. Simulation of the processes caused by a weak external perturbation of the system, resulting in an unstable state near the spinodal, may be especially fruitful.

7.2 Non-equilibrium States in Phase-Separating Binary Liquids and External Perturbations

One of the factors leading to non-equilibrium in binary phase-separating liquids is external perturbation with a variety of origins: electrical and gravitational fields, heating or cooling, and forces causing turbulence, such as shaking, stirring, flow generation, etc. The character and degree of efficiency of external perturbations are determined by their nature. If the efficiency of such an action on a binary system is sufficient to reach the region of absolute instability, a one-phase state arising as a result of continuous phase transitions evolves from the non-equilibrium state to a finite equilibrium state by the mechanism of spinodal decomposition.

This phenomenon was first investigated after fundamental work by J.W. Cahn and J.E. Hillard [187, 188], who in fact laid the foundations for the theory of spinodal decomposition. Their ideas were developed by J.S. Langer [184, 189–191], who suggested a more logical theory of spinodal decomposition based on a statistical approach to the problem. Spinodal decomposition in liquids was first found experimentally by J. Huang et al. [192]. Further, it was observed in many binary liquid systems including solutions of liquids with high molecular weights. The development of the spinodal decomposition in the late 1970s was rather fully described in a well-known review by V.P. Skripov and A.V. Skripov [185].

In recent years, many investigations have been made of spinodal decomposition in various systems. The range of substances under investigation has extended considerably, and methods of theoretical description and quantitative simulation have become more diverse [193–208]. There have been continued investigations in such traditional systems for studies on spinodal decomposition as alloys [209–214] and one-component systems with liquid–vapor phase separation [215–217]. Interesting results have been obtained for vitreous systems [218, 219] conditioned by the characteristics of the amorphous state. Many attempts have been made to modify the Cahn–Hillard–Langer theory [13, 200–203, 215, 220, 221]. It has also been suggested that a phase transition could be described within the framework of catastrophe theory [222], and the spinodal was shown to correspond to an assembly catastrophe.

A fractal theory of spinodal decomposition has been constructed. Theoretical analysis of morphological peculiarities in a structure where a phase transition occurs was carried out [223] using the spinodal decomposition model with conserved order parameter. Obtained in the linear approach, the solutions to the relaxation equation for expansion of the free energy in powers of the transformation parameter show that spinodal decomposition conditions a fractal structure in the formed phase, although the ordering does not give self-similar morphology. Spinodal decomposition in adiabatically closed systems considered in [224] is of practical interest. In the initial stages of spinodal decomposition, the latent heat of the transition and resultant change in temperature were shown to lead to significant changes in this process compared with those predicted by the Cahn–Hillard theory.

There is still much interest in the study of two-component systems. For example, the influence of hydrodynamic interactions on the coalescence of drops and roughness of system structure are discussed in [225]. The time dependence of order parameter fluctuations has been considered in the vicinity of the critical point of a liquid [226]. Properties of two-component spinodal systems [227] have been investigated, and in particular, the contribution of concentration fluctuations transferred by a shear hydrodynamical mode to the non-local diffusion coefficient determining the spinodal decomposition of binary liquids [228]. Other subjects of study have been intensity [229], temperature dependence [230], light scattering and small-angle scattering of X-rays [231] under spinodal decomposition for a non-equilibrium two-component mixture. The small value of diffusion coefficients for mixtures with high molecular weight compared to those with low molecular weight makes it possible to observe phase separation in detail [182, 232–239].

Of all the points investigated on the spinodal curve, the first was the critical point of phase separation. In order to describe it and its immediate vicinity, a reliable theory of critical phenomena was developed. Spinodal points far from the critical point were less definitely characterized by research results, using experimental data. Concepts analogous to the theory of second order phase transitions were worked out that gave rise to the idea of pseudo-critical phenomena. General principles for the latter are based on the Landau theory of self-consistent fields applied to describe thermodynamic properties of liquids near the spinodal. Since fluctuations in the spinodal region are as important as in the vicinity of a critical point, this tends to suggest identical behavior of the corresponding thermodynamic parameters, whose values comprise fluctuation contributions both at the critical point and at points on the spinodal curve. The 'pseudo-spinodal' hypothesis was put forward. This supposes a number of thermodynamic functions (e.g., the isochoric heat capacity function) in non-critical isochores to be of the same nature. The difference is that, in critical isochores, the divergence is observed at $T \to T_c$, whilst in non-critical ones, it is observed at $T \to T_H(\rho)$, where $T_H(\rho)$ is the 'pseudo-spinodal' temperature. Recent reports such as [240–242] have been devoted to investigations in this direction. Quite successful attempts have been made to apply the theory of renormalization groups, usually used in the analysis of critical phenomena, to spinodal decomposition [243–246].

Thus, the rising heat capacity observed in [104] falls within the framework of existing 'pseudo-spinodal' concepts and can be interpreted as a manifestation of spinodal decomposition in binary phase-separating liquids. On the other hand, the results of [104] can serve as an experimental confirmation of the existence of pseudo-critical phenomena.

In recent years, a new aspect has arisen in studies of spinodal decomposition of binary systems, namely the influence of an external action on the kinetics of decomposition, viz., mechanical perturbations such as stirring, shear flow, etc., or other external fields. There is, however, no single, accepted opinion as to how external perturbation factors influence spinodal decomposition. Therefore, any investigations into this problem are of great interest. Those studies that can be conventionally grouped according to the way they act are considered below.

7.2.1 Variable Electric Field

One of the first investigations in this direction was the work by Y. Yositaki and I. Akira [247]. They measured the intensity I of Rayleigh scattering (at angle 90°) of argon laser light due to concentration fluctuations in a binary cyclohexane–methanol mixture near the critical temperature T_c. An external variable electric field E was shown to cause a $\sim 65\%$ decrease in the scattering intensity I at $E \approx 700$ V/cm. A thermodynamic explanation for the delay and intensification processes affecting the formation of liquid nuclei in the vapor phase under an external electric field is given in [248].

7.2.2 Ultrasound

Manifestations of an external influence on the kinetics of spinodal decomposition are also known for solids. The influence of various kinds of ultrasound action on the kinetics of the oversaturated solid solution decomposition was considered in [249]. It has been established that the preliminary ultrasound processing of the investigated object stimulates the formation of a new phase. Introducing ultrasound into a sample directly in the process of artificial ageing retards decomposition of phases. Preliminary plastic deformation and ultrasound exposure accelerate the process of decomposition [250]. External tension not exceeding the fluidity limit significantly decelerates the separation of equilibrium phases. Thus, external mechanical and acoustic perturbations at various stages of separation affect the decomposition in different ways. In [251], the problems related to the dual influence of elastic energy on spinodal decomposition of a solid solution are considered within the framework of continuous and microscopic approaches.

7.2.3 Thermal Action

The system is transferred into a labile state by fast cooling (for solutions with an upper critical point of dissolution) or by fast heating (for solutions with a lower critical point of dissolution), whereupon spinodal decomposition occurs. The report [252] gives the results of light-scattering measurements in a critical binary iso-oil acid–water mixture under periodic changes in the temperature with respect to an average value slightly exceeding the critical temperature of the solution. The amplitude of temperature oscillations was such that the system periodically (with period 1 s) entered the labile region. Thermal quenching of the system can be considered as an external heat action. Hence investigations performed by thermal quenching with different intensities and at various regimes are of great interest. The processes of phase-separation via periodic cooling of liquid mixtures were tentatively investigated in [253]. Two cooling regimes were found. In one of them phase separation occurs, whilst in the other it does not. The temperature of the corresponding phase transition was determined. The structural factor was measured (by the light-scattering intensity) when maximal and minimal temperatures were reached. In the disordered state, such strong density fluctuations were found that the fluctuation–dissipation theorem was not applicable.

The influence of continuous quenching of the system at the initial stage of spinodal decomposition was considered in [254, 255]. Spinodal decomposition in a liquid that has been fast cooled in the region below a critical point was studied within the framework of the Langevin model [256]. The kinetics of the non-equilibrium phase transition of separation, together with the effect of initially existing concentration fluctuations on it, were investigated in [257]. The formation time of a modulated structure of spinodal decomposition was determined as a function of the rate of change of the thermodynamic state of the system in the lability region.

7.2.4 Optothermal Influence

One of the most interesting ways of inducing a labile state is the optothermodynamical method [258]. Using this method the previously reached rate of transfer of the solution through the mixing critical point was increased by a factor of 10^8–10^9 times, and its deep penetration into the region of lability was realized. As a result, in the early stages of coalescence the character of grain size growth in the heterogeneous phase was observed to deviate from temporal linearity. This is a consequence of solution 'memory' of the order parameter fluctuation in its initial thermostating in the homogeneous phase. Light-induced spinodal decomposition was studied by the optothermodynamical method in [259] in phase-separating solutions. This method makes it possible to get an unstable state in the bulk for relatively short times, inaccessible for many kinds of actions. However, it is not applicable for systems with an upper critical point of solubility and in investigations under isothermal conditions.

7.2.5 Noise Field

This is a factor affecting the fluctuation picture on the macroscopic level. The effects of the so-called coarsening transition on the dynamics of random phase interfaces induced by thermal noise were studied in [260]. The Allen–Cahn equations were used. These describe the evolution of an order parameter which is not conserved in time. It was shown that the noise tended to disorder the interfaces, which effectively decelerates the growth of domains, i.e., regions ordered in a certain way. A computation technique was suggested for calculating the temperature dependence of the deceleration coefficient. Two types of initial condition (completely random interface and 2-dimensional circulation domains) were considered.

The influence of random noise on spinodal decomposition was studied in [261, 262]. The method of cell dynamics was applied to simulations of spinodal decomposition in a two-dimensional system undergoing the influence of an external random excitement (noise). The method is based on solving the kinetic equation for the conserved order parameter in a variety of discrete space and time coordinates. The random excitement was described by the divergence of a vector with Gaussian components. The time evolution of the structure was characterized by a structural factor for which an average cluster size was obtained as a function of time. This function was proved to be a power law with exponent 0.28 or 0.33, depending on whether the noise is taken into account or not. The time of transition from one regime to another grew with increasing noise amplitude.

7.2.6 Turbulence

The most important kind of external influence is hydrodynamic perturbation since it can give rise to a redistribution of the substance concentration in a multicomponent system and allow one to observe decomposition in the isothermal regime (unlike thermal quenching). This determines its importance for possible technological applications. For binary solutions, turbulence was shown to mediate in solubility and phase-separation [263], as well as in nucleus formation.

When considering the influence of turbulence [264] on nucleus formation near a critical point of gas–liquid transition or a critical point of complete dissolution, it was found that:

- turbulence can delay the beginning of nucleus formation,
- it destroys large drops,
- growing drops coagulate fast in the turbulent flow.

The finite characteristic size of a drop depends to a large extent on the difference $T_c - T$ and the Reynolds number. Interesting results concerning liquid 'memory' effects revealed as a consequence of preliminary external perturbations (shaking) are shown in [265] for biological systems of high molecular weight. The turbulence effect in binary liquid mixtures is investigated in [263]. This effect also exists for critical phase-separating systems [267–270].

The processes of spinodal decomposition can be observed optically since the refractive index of developing heterogeneities can vary, and sufficiently large heterogeneities are revealed visually [271]. In [272] measurements of the extinction coefficient (EC) and light-scattering intensity (LSI) were made for critical mixtures 3-methylpentane–nitroethane and iso-oil acid–water in the presence of turbulence. The mixtures were observed to have no sharp interface between phases at temperatures a little below critical. The authors concluded that under such conditions there is no abrupt phase transition because turbulence suppresses long-wave fluctuations. However, through the temperature dependence of EC and ILS, they investigated the question of how one might determine the effective critical temperature. A hydrodynamical influence can reverse relaxation processes, as happened in the case of laser-light irradiation of a heterogeneous guaiacol–glycerine solution under mechanical perturbations [273]. A laser light beam that passed through the solution with a region of phase separation at a temperature slightly shifted towards the heterogeneous region caused a light ring to form on the screen due to light refraction on drops formed during the process. When shaking the dish the solution again became homogeneous, and then drop-formation was repeated once more.

7.2.7 Shear Flow

A large number of research projects have been carried out under conditions of shear flow. Some experiments and theoretical descriptions of various physical phenomena near critical points, where relatively small perturbations can bring the system into a highly non-equilibrium state, are reviewed in [274]. Critical fluctuations under so-called strong shear, when the velocity of flow with shear is much larger than the reciprocal lifetime of these fluctuations, spinodal decomposition of a mixture under shear, nucleus formation under weak shear, and turbulence in critical binary mixtures were studied there. Other effects of this kind, in particular the effects in polymers and liquid crystals, are mentioned briefly in this report.

The kinetics and morphology of the spinodal decomposition of a liquid under conditions of hydrodynamical flow are considered in [275]. A mixture of isobutyl acid with water at critical concentration was located in a Kuept cell consisting of two coaxial cylinders. Decomposition of the mixture was observed using He-Ne laser light scattering for various degrees of overcooling. It was concluded that hydrodynamical flow influences the character of the forming domain structure if the rate of shear is larger than the domain growth rate. Under these conditions, domain growth becomes anisotropic and is characterized by specific scaling properties.

By making use of numerical simulations of two-dimensional phase transitions in ordinary shear flow, the influence of hydrodynamics on the growth of domains was studied in [276]. Simulations were carried out using a conserved moment model of the gas–liquid system for a binary immiscible liquid. It was shown that anisotropic anomalies observed in tentatively obtained structural functions were caused by the induced shear of smectic-type domain ordering.

It has been established that the coexistence curve of binary [277] and simple [278] liquids shifts under shear flow. Anisotropy in light absorption [279] and birefringence [280, 281] arise at critical solution concentrations. Other aspects of the way shear flow influences unstable states are considered in [282–285].

7.2.8 Centrifugal Forces

By the nature of their action, perturbations induced by a centrifugal field are similar to those induced by shear flow. Experimental studies of the influence of centrifugal forces on static and dynamical properties of the system nitrobenzol–n-hexane near a critical point where these liquids mix are described in [286]. The distribution of the refractive index gradient dn/dy (and hence also the distribution of the concentration gradient dx/dy) was measured by photographic methods in the mixture as a function of the distance y to the rotation axis of the dish at different times t and at various angular velocities ω_y (from 3×10^3 to 3×10^4 rev/s) in the temperature range $T - T_c = 0.5$–10 K. The data obtained were used to determine the susceptibility $dx/d\mu$ in a wide range of deviations of the concentration x from its critical value, as well as the sedimentation coefficient S of the mixture. The value of S was found to increase significantly near a critical point: $S \sim (T - T_c)^{-1 \pm 0.1}$.

The results of the experiment are in a good agreement with the theory of static and dynamic similarity. During the experiment, at very large values of ω_y, some non-linear effects, including mixture instability near T_c, were also observed.

7.2.9 Stirring

Stirring is the most frequently investigated type of hydrodynamic perturbation. One of the first studies describing the non-trivial influence of stirring was [287]. Temperature heterogeneities appearing during stirring were revealed there. The effect of stirring was studied for critical systems in [288,289]. In experiments on light scattering [290], a strong suppression of phase separation was observed near a critical point under mechanical perturbation of a binary liquid mixture. Stirring performed at values of the Reynolds number from $R = 6 \times 10^9$ to 4.5×10^4 resulted in a decrease in effective critical temperature from ~ 1 mK to 50 mK. A simple model has been suggested to explain this phenomenon, based on suppression of the concentration fluctuation by viscous effects.

The influence of turbulent stirring in a binary liquid mixture at temperatures lower than critical has been investigated in [291], when phase separation of a mixture occurs in the absence of external forces. The linear theory of stability of concentration fluctuations shows that turbulent stirring successfully competes with phase-separation processes. However, upon further cooling of the mixture, the conditions arise for growth of concentration fluctuations with wavelengths from the viscous–convective range of the spectrum.

In [292], critical phenomena in binary mixtures are considered under remotely operated passive stirring. The continuous model was used, assuming that the fluctuating diffusion flux and random velocity field are described by Gaussian functions, that the liquid is incompressible and that the time scale of velocity fluctuations is much less than that of concentration fluctuations. By the renormalization group method and ϵ-expansion for critical indices with respect to the dimension $d = 4$, the exponent k in the law $\Delta T_c \to R^k$ defining the dependence shift (decrease) of the critical temperature on the Reynolds number R was found to be 1.74.

In [293], where time scales of domain formation characterizing interactions in a stirred mixture of liquids were considered, it was noted that if a system with two phases of the liquid mixture at supercritical temperature $T > T_c$ was being intensively stirred, then the time of their mutual dissolution increased considerably. The authors gave the following explanation: stirring caused cooling of the mixture, decreasing the temperature to precritical $(T < T_c)$. Subsequent heating of the system via interaction with a reservoir at supercritical temperature increased the temperature to its supercritical value and led to dissolution of the phases.

The time correlation function $g(t) = \langle I(0)T(t)\rangle$ of the light intensity scattered by a binary mixture of liquids near a critical point was measured

7.2 Non-equilibrium States in Phase-Separating Binary Liquids

in [294]. The liquid was under continuous turbulent stirring with a frequency of $\omega = 5$–15 Hz that corresponds to the Reynolds numbers $\sim 10^3$. The form of the function $g(t)$ was found to change completely when varying the local velocity gradient (turbulent shear) from weak values to strong. At strong shear, the function $g(t)$ became highly exponential with decreasing index when approaching a critical point. This behavior was associated with suppression of critical fluctuations due to the turbulent stirring.

A review of investigations of new, non-equilibrium effects in mixable liquids has been given in a paper by A. Onuki [295]. The investigation focused on the idea that stirring has an effect on thermodynamic properties of simple liquids. Stirring was assumed to be conditioned by external factors, and emergent flows could be of any type: laminar, turbulent or mixed. Special attention was drawn to the influence of stirring on critical behavior and phase separation. The author found that stirring of liquids in the gravitational field gave rise to a vertical temperature gradient. Critical behavior of specific heat capacities was considered. Experimental data on dynamical light scattering in turbulent binary mixtures were analyzed. The paper showed that stirring significantly influenced nucleus formation. To explain the revealed behavior of liquids, a theoretical model was suggested and the results were analyzed.

It is clear that many data in the above-mentioned works were obtained in the range of near-critical concentrations. As is well known, in the very close neighborhood of a critical point, the distance between the binodal and spinodal is very small, so that it is difficult to distinguish experimentally effects related only to transition of the system through spinodal from a labile region to a metastable region. In fact, it is possible to observe the transition from an unstable state into a stable state, omitting the metastable state. For this reason, it is of interest to carry out investigations at concentrations far from critical, as was done in [104].

The analysis in the literature attests to the decelerating effects of external stirring on spinodal decomposition in phase-separating binary liquids, though many details are still lacking. For this reason, it is primarily the optical characteristics of systems under stirring that were investigated. The problems related to changes in thermophysical parameters and the structure of the system itself due to external actions have been neglected by researchers.

It is of special interest to consider the way very weak perturbations influence non-equilibrium processes in phase-separating liquids. This influence would not a priori appear to be of any significance. The unexpected behavior of the heat capacity for sufficiently weak actions (mechanical stirring at frequencies up to 1 Hz) demonstrates, however, that even for such low levels of perturbation, qualitatively new states of the system are reached. The transition to such states looks like a non-equilibrium phase transition [104].

7.3 External Perturbation and Spinodal Decomposition

Let us choose weak mechanical stirring as the perturbation, and the interface of weakly solubilized liquids as the subject of investigation. A two-component phase-separating liquid system is convenient for study by stationary methods. On the one hand, the dynamics of the structural reconstruction is determined by diffusion, and on the other hand, such a system has a high degree of susceptibility that can affect the processes of excitation relaxation.

By taking into account the above-mentioned facts, the specific heat capacity was measured by the method of adiabatic calorimetry in a phase-separating system of methanol–n-heptane. Measurements were conducted under stirring generated by the motion of stirrers caused by a periodic magnetic field at a frequency of 0.06–1 Hz [104].

The heat capacity was measured as follows. To begin with, a temperature was set in the cell that was lower than the temperature of the one-phase states. Then the calorimeter heater and electromagnetic mixer were simultaneously turned on. After supplying a certain amount of heat, the heater was turned off, stirring continued for a while, and then the sample temperature was measured. A one-fold interchange of the regimes with heating and without it formed an elementary step in the temperature run. To separate a possible contribution of the stirring, the measurements were also conducted without a mixer. In this case, after stopping the heating, the system was settled for a while so as to balance the temperature gradient of the system, and then the temperature was registered.

Figure 7.4a and b shows the measurements of the dependence of isochoric heat capacity C_V on temperature T at various stirring rates. We see that with increasing stirring intensity, the anomalous behavior of the heat capacity at temperature 264.1 K with changing amplitude (Fig. 7.4b) becomes more definite. The peak of the heat capacity at 303.6 K has a constant amplitude and corresponds to the disappearance point for visual observations of the meniscus. Such behavior of the heat capacity points to a significant change in the system state near 264.1 K. This corresponds to its leaving the region of absolute instability, i.e., moving out of the spinodal. Thus, as a result of weak mechanical stirring, the investigated system is transferred from an initially stable state into the region of absolute instability under the spinodal.

This is confirmed by the experimental coexistence curve constructed from the heat capacity data (Fig. 7.5). Curve 2 is similar to the spinodal, confining the region of metastable and absolutely unstable states, and satisfactorily described by the expression

$$\frac{x' - x''}{x_c} = B_{\mathrm{Sp}} \left[\frac{T_{\mathrm{Sp}} - T_c}{T_c} \right]^{\beta_{\mathrm{Sp}}},$$

where x' and x'' are concentrations of separating phases, B_{Sp} is the critical amplitude, and T_{Sp} is the spinodal temperature. For the methanol–n-heptane system, $B_{\mathrm{Sp}} = 4.2 \pm 0.2$, $\beta_{\mathrm{Sp}} = 0.4 \pm 0.05$.

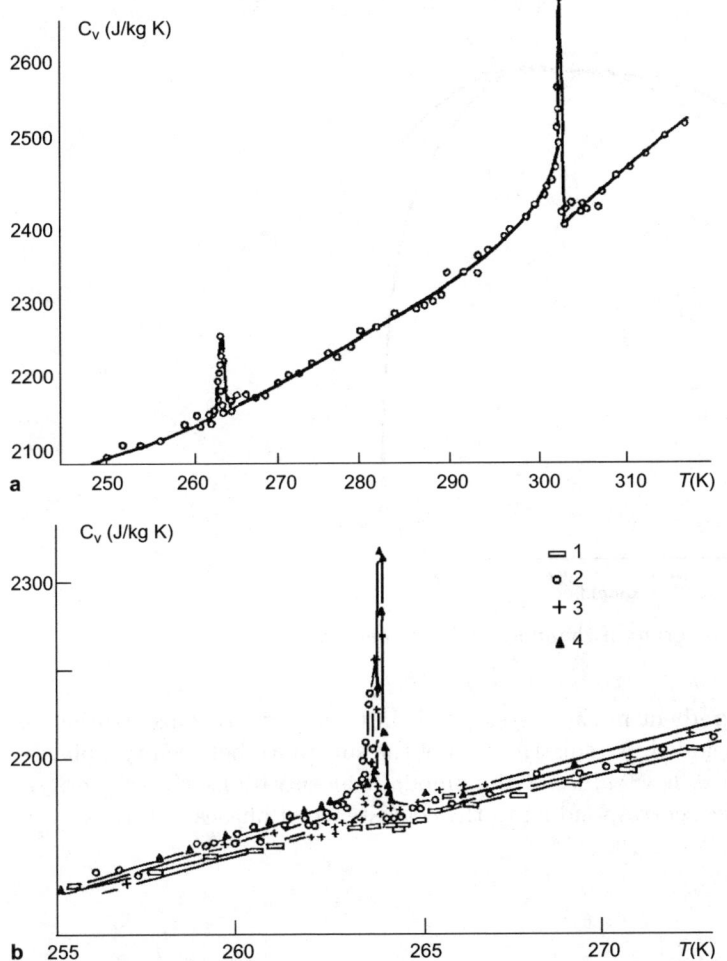

Fig. 7.4. Temperature dependence of isochoric heat capacity (**a**) and amplitude of the first heat capacity jump (**b**) at various stirring frequencies: (curve 1) 0, (curve 2) 0.016, (curve 3) 0.1, (curve 4) 1.0 Hz

Let us imagine what processes can occur in the system on the phase interface with mixed concentration x_0 determined by the initial experimental conditions.

7.3.1 Heating of a System Without Stirring

After every discrete introduction of heat, the system temperature increases by δT, and correspondingly concentrations of coexisting phases (x_{B1}, x_{B2}) fail to satisfy the equation of chemical potentials. This results in a non-equilibrium

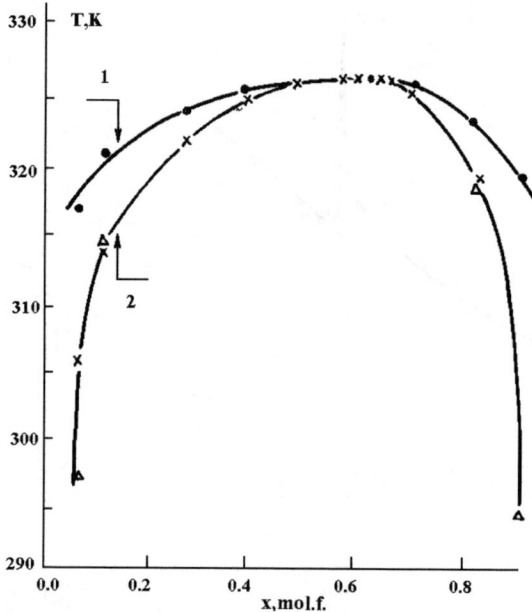

Fig. 7.5. Phase diagram of the methanol–heptane system

concentration gradient in the system. This in turn leads to concentration relaxation accompanied by redistribution of the substance between two phases. As a consequence, new values of the binodal concentrations x'_{B1}, x'_{B2} are set in the two phases, corresponding to the coexistence of phases at temperature $T + \delta T$ (Fig. 7.6).

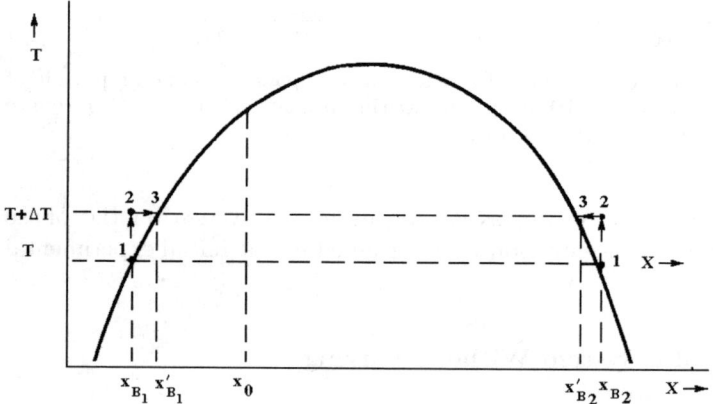

Fig. 7.6. Coexistence curve of a binary mixture under heating of the system without stirring

Thus, in the absence of other factors of external action, at each instant of time the system has binodal points (points 1 and 3) or the closest in value overbinodal concentrations (point 2). Therefore, under such conditions it is not possible to obtain recurring values of the heat capacity related to metastable states, let alone labile states.

7.3.2 Heating of a System by Stirring

In the state chosen initially, stirring itself does not form a one-phase system. By durable and energetic stirring only an unstable emulsion can be obtained. This then gradually and spontaneously separates into two liquid phases. In the case of weak stirring, such a state cannot be obtained. Although stirring tends to homogenize the whole system by attracting the concentration values of heterogeneous phases to the average x_0, the diffusion instability does not lead to interdissolution of phases. However, the effect of weak stirring can manifest itself in local regions near the phase-separation interface, where sensitivity to external perturbations is a little higher than in the rest of the system.

In equilibrium, in the interface region of weakly solubilized liquids within the correlation length, the system state is always realized with average concentration x_0 of a mixed state corresponding to absolute instability. However, phase separation is absent. With an increase in the system temperature, the chemical potentials of phases change and become unequal in value. As a consequence, non-compensated flows of the substance emerge that in the end give rise to a new, stable heterogeneous state. During non-equilibrium mass transfer, the concentration gradient near the boundary becomes more fluent and the width of the region of unstable concentrations increases. This region is in transient dynamical equilibrium and highly sensitive to perturbing factors, in particular, to stirring. Stirrers periodically cross the interface of phase separation, changing the concentration gradient in a random way, and thereby producing composition fluctuations, whilst dragging small regions of one phase towards another. As a result, stirring causes dispersion of one phase within another in the vicinity of the phase-separation interface.

Clearly, weak stirring cannot form any extensive macroscopic homogeneous state with unstable concentration. However, some microzones with metastable and labile concentrations located in the region of the phase-separation interface can arise under the action of stirring, due to viscous and hydrodynamical effects. The region including microzones becomes rather transient in equilibrium, i.e., the region with mixed concentration x_0 is effectively extended. Labile microzones are unstable with respect to any fluctuations and decay spontaneously, whereupon local spinodal decomposition takes place. Metastable microzones are a little more stable and some closely adjoining microzones can even coexist. However, to begin with, stirring generates new microzones, and subsequently decelerates the decomposition of existing microzones. External stirring thus influences the level of fluctuations and induces a non-equilibrium state in the system. When approaching the temperature which is spinodal for x_0, the number of labile unstable zones increases because of the decrease in the difference between stable and labile concentrations $\delta x = x_\mathrm{B} - x_\mathrm{S}$.

In the region under the spinodal, $(\partial P/\partial V)_{T=\mathrm{Const.}} > 0$. This leads to the instability of isothermal processes with respect to small composition fluctuations, because the decrease in volume implies compression of the system and a reduction in pressure, which gives rise to further compression. On the spinodal curve, $(\partial P/\partial V)_T = 0$. Hence, in the region of pseudo-criticality, $(\partial P/\partial V)_{T=\mathrm{Const.}}$ reverses its sign, whereas the quantity $(\partial P/\partial V)_{S=\mathrm{Const.}}$ remains negative [186]. Therefore, adiabatic compression of some volume elements and extension of others, contrary to isothermal processes, leads to an increase in pressure in compressed regions of the volume and to its decrease in solubilizing regions. This in turn means a reversed equalization of changes in composition in an unstable one-phase state. Thus, local changes in the system state require exchange of heat and movement of the substance between extended and compressed regions of the volume. For this reason, separation into two phases is related to heat capacity and diffusion. Under fast pressure equalization, in the course of decomposition of the system state under the binodal, its instability is expressed via the emergence of a heat capacity jump at fixed pressure, as given by the relation

$$\frac{C_P}{C_V} = \frac{(\partial P/\partial V)_S}{(\partial P/\partial V)_T}.$$

The sign reversal of $(\partial P/\partial V)_T$, when passing through zero, causes a sign reversal of C_P when passing through $\pm\infty$.

Following the above discussion, one can conclude that weak stirring creates a local non-equilibrium state, where spinodal decomposition occurs. The non-equilibrium state changes its structure in a way that is revealed as a non-equilibrium phase transition.

The influence of stirring can be regarded as a thermodynamic process where the system state function changes under the action of external forces. Let us consider the contribution of stirring to the free energy. An external perturbation can be defined as the work done on the system. Then the total free energy F of the system consists of the sum of the free energy F_0 without taking into account the stirring and the contribution of perturbation ω, i.e.,

$$F = F_0 \Omega, \tag{7.4}$$

where

$$F_0 = \int \left[f(x) + K(\nabla x)^2 \right] \mathrm{d}V, \quad \Omega = \int H(\omega, x) \mathrm{d}V,$$

and $f(x)$, $H(\omega, x)$ are the free energy density of a homogeneous solution and the energetic contribution of the perturbation, respectively.

The efficiency of stirring is higher for local concentrations which are further removed from x_0. If anywhere in the bulk the local concentration

$x(r) = x_0$, the stirring action does not manifest itself because of the closeness of the local concentrations values to x_0 under perturbation (homogenization effect of stirring). Let us consider small deviations of $x(r)$. By expanding $H(\omega, x)$ as a Taylor series in x near x_0, we obtain

$$H(\omega, x) = H(\omega, x_0) + \frac{\partial H}{\partial x_0}(x - x_0) + \frac{1}{2}\frac{\partial^2 H}{\partial x_0^2}(x - x_0)^2 + \dots .$$

As mentioned, stirring does not influence the substance distribution when the local concentration is already equal to the average value. This implies that the first term of the expansion is zero. From the normalization condition

$$\int (x - x_0) dV = 0 ,$$

we find that the linear term disappears. Taking into account the smallness of the quantity $x - x_0$, we limit ourselves to the quadratic term:

$$H(\omega, x) = h(\omega)(x - x_0)^2 , \quad h(\omega) = \frac{\partial^2 H}{\partial x_0^2} . \tag{7.5}$$

Substituting (7.5) into (7.4), we obtain for the total free energy of the solution

$$F = \int \left[f(x) + K(\nabla x)^2 - h(x - x_0) \right] dV . \tag{7.6}$$

Note that a form of the free energy very similar to (7.6) was found for systems of solid solutions, where composition fluctuations accompanied by elastic tensions are taken into account [185]. In that case,

$$F = \int \left[f(x) + K(\nabla x)^2 + \xi^2 Y(x - x_0) \right] dV ,$$

where ξ is a relative change in the lattice parameter when changing the composition, and Y is a parameter determined by elastic constants. Later, we will obtain the dependence of F on stirring in a more accurate way.

7.4 Statistical Account of an External Stirring Field

Let us consider a two-component separating system (weakly solubilized liquids) of volume V_0 and let x be the relative concentration of a specific component. For the spatially non-equilibrium distribution, $x(r)$ is a function of a concrete point of space r with some average concentration,

$$x_0 = \frac{1}{V_0} \int x(r) \, dV . \tag{7.7}$$

Let us suppose stirring has resulted in system conditions for which a one-phase state is unstable and various concentration heterogeneities are not damped. Then the function $x(r)$ will take an arbitrary (stochastic) form (Fig. 7.7a) satisfying (7.7). In order to consider a mechanism of external

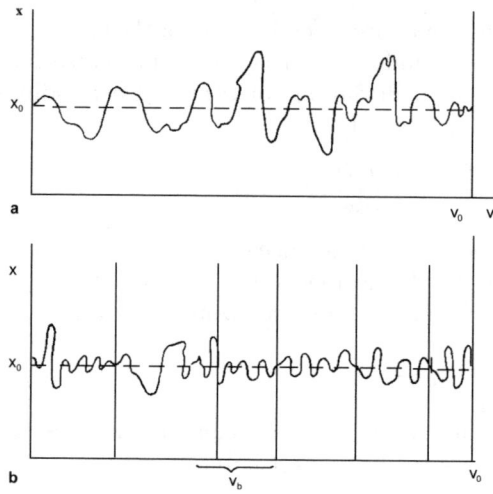

Fig. 7.7. Concentration distribution curves in the bulk: (**a**) stochastic (disordered) state, (**b**) block state

action, we introduce a model for the system that is similar to the Langer model [58]. We divide the system into small cells of volume $V_i = N_i d^3$, where N_i is the number of particles in a cell, d is the interparticle distance, and i is the cell number. For each cell, a corresponding relative concentration x_i exists, determined as an average value per cell:

$$x_i = \frac{1}{V_i} \int_V x(r) \, dV \; .$$

In order to consider x_i as a continuous value, a sufficiently large number of particles is needed in the cell. Consequently, the volume V_i is limited from below by the condition $V_i \gg d$. On the other hand, in a continuous description employing the quantity x_i instead of $x(r)$, the upper bound $V_i \ll V_0$ is also needed. Thus, the choice of volume V_0 is limited by the condition $d \ll V_i \ll V_0$.

In the general case, mechanical stirring deflects a system from equilibrium, bringing about coupling of adjacent elements in the volume on account of mass transfer, and contributing correlations on a macroscopic scale. In this case, local fluctuations become non-Poissonian due to the fact that non-equilibrium is maintained. Since fluctuations are determined by local volume, one can choose a level V_i in the system such that fluctuations remain Poissonian. Regarding cells as such regions, we suppose that stirring serves to shuffle the cells among each other. The result of this coarse-grained stirring may be a large-scale homogeneity where values of average concentrations of some micro-regions will be close to each other.

7.4 Statistical Account of an External Stirring Field

The system can thus be divided into a number of blocks of volume V_i (Fig. 7.7b). The average concentration over a block is x_0:

$$x_i = \frac{1}{V}\int_V x(r)\,dV = \frac{1}{n}\sum_{i=m}^{bn} x_i \approx x_0, \qquad (7.8)$$

where $n = V_i/V_0$ and $M = (b-1)n + 1$. The larger the stirring intensity, the higher the degree of homogeneity of the system. Accordingly, the sizes of blocks become smaller and smaller. In the asymptotic limit of very intensive stirring, one can assume that $V_b \to V_i$.

A set of values of x_i constitutes a certain state configuration $\{x\}$ for which one can calculate the energy $E\{x\}$ and statistical weight $W\{x\}$. Let us introduce a statistical sum of the system for a certain x_0:

$$Z(x_0) = \prod_{i=1}^{m}\int dx_0 W\{x\} e^{Ex/k_B T}\delta\left(\frac{1}{m}\sum_{i=1}^{m} x_i - x_0\right) Q(x_0), \qquad (7.9)$$

where $m = V_0/V_i$. The δ function serves to discard those configurations $\{x\}$ that do not satisfy the condition of conservation of particle number in the system, and the function $Q(x_0)$ implies a selection of configurations corresponding to block states. So the quantity $Q(x_0)$ can be expressed in the form

$$Q(x_0) = \prod_{b=1}^{Y-1}\delta(x_b - x_{b+1}), \quad Y = \frac{V_0}{V_b}, \qquad (7.10)$$

where Y is the number of blocks in the system.

Since (7.8) is satisfied in the block state, the average concentration for a block is equal to the average concentration of the rest of the system of $Y-1$ blocks. Therefore the function δ in (7.10) can be rewritten

$$\delta(x_b - x_{b+1}) = \delta\left[x_b - \frac{1}{Y-1}\sum_{b-1}^{Y} x_b(1-\delta_b)\right]$$

$$= \delta\left[\left(1 + \frac{1}{Y-1}\right)x_b - \frac{Y}{Y-1}x_0\right].$$

For larger Y we make the approximation

$$\delta(x_b - x_{b+1}) = \delta(x_b - x_0).$$

Moreover, x_b can be identified with some local concentration $x(r)$, whose coordinate r corresponds to the coordinates of the center x_b. Substituting (7.10) into (7.9), we find finally

$$Z(x_0) = \prod_{i=1}^{m}\int dx_0 W\{x\} e^{E(x)/K_B T}\delta\left(\frac{1}{m}\sum_{i=1}^{m} x_i - x_0\right)\delta(x - x_0). \qquad (7.11)$$

In subsequent analysis it will be convenient to use the following approximation for the δ function:

$$\delta(x - x_0) \approx \frac{\exp\left[\dfrac{(x - x_0)^2}{q}\right]}{\sqrt{\pi q}}, \tag{7.12}$$

where $q \sim (\Delta x)^2$ is the degree of smearing of the statistical picture, assumed to be sufficiently small. By taking into account (7.12) and assuming $F = k_B T \ln Z$ from (7.11) for rough cellular division, we obtain the expression for the free energy,

$$F\{x\} = E\{x\} + \frac{k_B T}{d}(x - x_0)^2 - k_B T \ln \frac{W(x)}{\sqrt{\pi q}}. \tag{7.13}$$

The work done by external forces makes a negative contribution to F, and the field component in (7.13) must have a negative sign. When considering a continuous picture, where energy densities are introduced, the relation (7.13) taking into account the fluctuation contribution will have the form

$$F = \int \left[f(x) + K(\nabla x)^2 - h(x - x_0)^2 \right] dV, \tag{7.14}$$

where h is a generalized external field related to stirring of the solution. Thus, to a rough approximation, the influence of external stirring makes a quadratic contribution in $x - x_0$. The dependence of the free energy on an external perturbation turns out to be the same as we assumed in (7.6).

Let us consider the character of changes in the system entropy after turning on an external stirring field. Let $g_0(x)$ and $g_b(x)$ be distribution functions for the concentrations for the whole system and for a block, respectively, satisfying the normalization conditions

$$\int g_0(x) dx = N, \quad \int g_b(x) dx = \frac{N}{Y},$$

where N is the total number of particles of the separate component. Inserting some function $g^*(x)$ with the normalization condition

$$\int g^*(x) dx = 1, \tag{7.15}$$

we obtain

$$g_0(x) = N g^*(x), \quad g_b(x) = g^*(x) \frac{N}{Y}. \tag{7.16}$$

As is known from [296], the entropy of a statistical system is determined as

$$S = -k \int z(r) \ln z(r) \, dr,$$

where $z(r)$ is the probability distribution function of the system location in the parameter interval $(r, r+dr)$. We determine the system entropy S_0 before stirring and S_1 after stirring

7.4 Statistical Account of an External Stirring Field

$$S_0 = -k \int g_0(x) \ln g_0(x) \, dx , \tag{7.17}$$

$$S_1 = \sum_{b=1}^{Y} S_b = Y S_b , \quad S_b = -k \int g_b(x) \ln g_b(x) \, dx ,$$

where S_b is the entropy of a block. Let us consider a change in the entropy due to external action,

$$\Delta S = S_1 - S_0 . \tag{7.18}$$

Using (7.17), we get

$$\Delta S = -k \int \left[g_b(x) \ln g_b(x) - g_0(x) \ln g_0(x) \right] dx .$$

From (7.16) it follows that

$$\Delta S = -k \int \left[g^*(x) \ln \left(\frac{N}{Y} g^*(x) \right) - g^*(x) \ln \left(NY g^*(x) \right) \right] dx$$

$$= kN \ln Y \int g^*(x) \, dx .$$

Thus, from the normalization condition (7.15), we obtain finally

$$\Delta S = kN \ln Y . \tag{7.19}$$

As noted, more intensive stirring decreases the volume V_b of the blocks and increases Y. On the other hand, in the absence of external influences, there always exists one large block, the whole system. Therefore, the dependence of Y on the stirring frequency ω can be expressed as

$$Y = 1 + Y(\omega) .$$

For small stirring frequencies ω, (7.19) takes the form

$$\Delta S = kNY(\omega) . \tag{7.20}$$

If the definition of the heat capacity $C_V = T(\partial S/\partial T)_V$ is rewritten in a rough approximation

$$C_V \approx T_0 \frac{\Delta S}{\Delta T} ,$$

where ΔC_V is the amplitude of the jump in C_V, T_0 is the temperature in the immediate vicinity of which the irregular behavior of C_V is observed, and ΔT is the temperature interval of this irregularity, then we obtain from (7.20)

$$\Delta C_V \sim Y(\omega) . \tag{7.21}$$

Therefore, since the observed jump in the heat capacity in the methanol–n-heptane mixture of weakly soluble liquids is proportional to the degree of homogenization, this tentatively confirms the formation of a relaxation phase as a result of weak stirring on the boundary of these liquids [104].

8. Weak Stirring and Absolute Instability Phenomena

8.1 Singularity in the Heat Capacity in Forced Spinodal Decomposition

In the experiments we have carried out, two jumps were observed in the heat capacity of binary methanol–n-heptane mixture with weak stirring of the solution. These were registered at points of the spinodal and binodal curves on the diagram of states, and the magnitude of the jumps depended on the stirring frequency. The authors of [104] considered their appearance to be connected with spinodal decomposition, although initially the whole system was beyond the spinodal.

Let us consider the influence of weak stirring on fluctuations in the local concentration of the methanol–n-heptane mixture according to the mean field approach. Regarding the solution concentration $x(r)$ as a random function of coordinates, we use the free energy dependence on the concentration in the form (7.14). The change in the free energy of a homogeneous state with respect to the perturbed one is determined as

$$\Delta F = \int \left[\frac{1}{2} \frac{\partial^2 f}{\partial x^2} \delta x^2(r) + K \left(\nabla \delta x(r) \right)^2 - h \delta x^2(r) \right] dV . \tag{8.1}$$

If the energy contribution of the mixer is considered to be mainly determined by the resistance force of the liquid to the oscillatory movement of the stirring centers (spheres of radius R), the force acting on the stirring centers is determined by the formula [297]

$$f(t) = \eta G_1 u + G_2 \left(\frac{2\eta \rho}{\omega} \right)^{1/2} \frac{du}{dt} ,$$

where η and ρ are the dynamical viscosity and density of the binary liquid, ω is the frequency of oscillation of the stirring centers, u is the velocity of their movement, and G_1, G_2 are geometrical parameters of the system. Supposing that

$$u = u_0 \cos(\omega t) , \quad \frac{du}{dt} = -a_0 \sin(\omega t) ,$$

where u_0, a_0 are amplitudes of the velocity and acceleration of a stirring center, we can find the force averaged over half an oscillation period of the centers, viz.,

$$\bar{f} = \frac{2}{w} \int_0^w f(t)\,\mathrm{d}t = 4G_2 a_0 \left(\frac{2\eta\rho}{\omega}\right)^{1/2},$$

where W is the oscillation period of the centers. The contribution of an external field h at small frequencies is determined by the relation

$$h = \zeta \left(\frac{1}{\omega}\right)^{1/2},$$

where ζ is a coefficient accounting for the effect of the system geometry in energy dissipation

$$\zeta \approx \text{Const.} \times R^2 a_0 l (2\eta\rho)^{1/2},$$

and l is the size of a cell.

As the analysis shows, a homogeneous state obtained from (8.1) is stable with respect to the formation of small concentration heterogeneities [185] for the correlation radius r_c given by

$$r_c = \frac{2\pi}{\left\{-\frac{1}{2K}\left[\frac{\mathrm{d}^2 f}{\mathrm{d}x_0^2} - \zeta\left(\frac{1}{\omega}\right)^{1/2}\right]\right\}^{1/2}}. \tag{8.2}$$

Hence, the characteristic scope of fluctuations depends on the stirring frequency. The growth of mean-square fluctuations, i.e., the increase in r_c, is equivalent to the system moving inside the spinodal region, because beyond it, the fluctuations disappear due to diffusion, and the characteristic spatial scale r_c of heterogeneities decreases. Thus, at a fixed temperature, growing fluctuations give rise to sections in the phase diagram where the local concentration looks as though it has been 'dragged' under the spinodal [298]. In its turn, the growth of mean-square fluctuations contributes to a jump in the heat capacity [78].

It is possible to evaluate the scale of the mean-square concentration fluctuations of the solution within the framework of the self-consistent field theory at small values $h(\omega)$ of the external field, $x(r)$ diverging weakly from x_0. Then, according to the condition of minimal thermodynamic potential (8.1), the equation for $\delta x(r)$ is linear to a first approximation:

$$\frac{\delta F}{\delta x(r)} = -K\Delta\delta x(r) + \frac{\mathrm{d}^2 f}{\mathrm{d}x^2}\delta x(r) - h\delta x(r) = 0. \tag{8.3}$$

The corresponding solution of (8.3) will be the Green function

$$G(r) = \frac{1}{4\pi K r} \exp\left(-\frac{r}{r_c}\right).$$

As shown in [78], $G(r)$ coincides with the concentration fluctuation $\langle \delta x(r_1) \delta x(r_2) \rangle$ up to a multiplicative factor. Therefore, the fluctuation contribution to the heat capacity expressed via $G(r)$ depends on the external field $h(\omega)$ via $r_c(\omega)$, viz.,

8.1 Singularity in the Heat Capacity in Forced Spinodal Decomposition

$$\Delta C = \frac{2a}{T_c} \lim_{r \to 0} \frac{\partial G(r)}{\partial T} \approx \frac{a^2 \omega^{1/4}}{8\pi T_c^2 K^{1/2} \zeta^{1/2}} \, . \tag{8.4}$$

Taking into account the fact that

$$\lambda_c \approx \frac{bT_c^{1/2}}{(T_c - T)^{1/2}} \approx 0.001 \text{ cm} \, , \quad T_c - T \approx 0.01 \text{ K} \, ,$$

we get $b \approx 10^{-7}$ cm, $a \approx 0.1$ [78], $G(r) \approx \exp(-r/r_c)$. Since $l \gg r_c$, the fluctuation added to the heat capacity at $\omega = 1$ Hz in (8.4) is 0.1 of the absolute value corresponding to the experimental value found in [104]. Figure 8.1 shows the frequency dependence of the heat capacity jump,

$$\frac{\Delta C(\omega)}{\Delta C(\omega = 1)} = \omega^{1/4} \, .$$

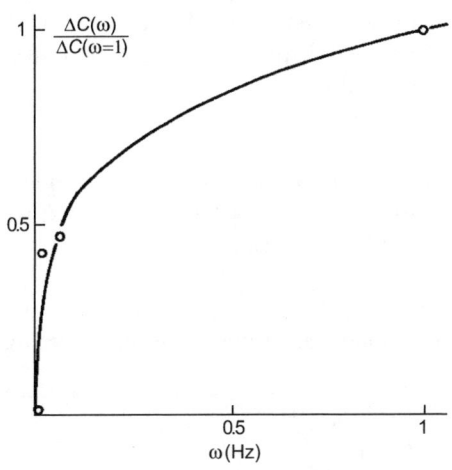

Fig. 8.1. Frequency dependence of the magnitude of the jump in heat capacity. *Circles* denote data from [104] and the *continuous curve* is a plot of (8.4)

Thus, despite the fact that a system lies beyond the spinodal 'on the average' under weak stirring, and hence that there are sections of the phase diagram in the region of stability where the local concentration proves to be inside the spinodal, the sizes of such sections and the penetration depth under the spinodal depend on the stirring frequency ω. This in turn is attested by the growth of the jump in the heat capacity with the growth of ω along lines determined by the conditions

$$\frac{df}{dx} = 0 \quad \text{and} \quad \frac{d^2 f}{dx^2} = 0 \, .$$

It is also attested by the separating regions of the concentration equalization with respect to volume and the spatial scale increase of their fluctuations. The structure formed at $\omega = 0$ can separate for a short time after binodal decomposition.

8.2 Extending the Region of Absolute Instability

The process of spinodal decomposition is described by the non-linear Cahn–Hillard equation (7.1). To consider the influence of external stirring, one must introduce an analogous diffusion equation via the expression (7.14) for the free energy taking into account external action:

$$\frac{\partial x}{\partial t} = L\nabla^2 \left[\frac{\partial f}{\partial x} - 2K\nabla^2 x - h(x - x_0)\right] . \tag{8.5}$$

For our purposes, it is convenient to make use of the free energy density of a homogeneous solution in the form considered in [300]:

$$f(x) = \varepsilon_0 \left[-(x_{S2} - x_{S1})^2(x - x^*)^2 + \frac{2}{3}(x - x^*)^4\right] , \tag{8.6}$$

where $x^* \equiv (x_{S1} + x_{S2})/2$, x_{S1} and x_{S2} are defined as points at which the second derivative of the free energy density with respect to the concentration tends to zero. By substituting (8.6) into (8.5) and introducing the deviation $\check{x}(r,t)$ of the concentration from its average value x_0,

$$\check{x}(r,t) \equiv x(r,t) - x_0 ,$$

one can obtain the non-linear Cahn–Hillard equation with stirring effects included. However, this is cumbersome, so we limit ourselves to the simpler and at the same time important case of small fluctuations,

$$\check{x} \ll x_0 .$$

Let us change to dimensionless time and space coordinates by introducing characteristic scales [300, 301]:

$$t' = \frac{a^2}{Z} , \quad \text{where} \quad Z = u\varepsilon_0 , \quad y' = a \equiv \left(\frac{K}{\varepsilon_0}\right)^{1/2} .$$

Working to third order, we get

$$\frac{\partial \check{x}}{\partial t} = 8x_0 \left[\left(\Delta x_0 - \frac{1}{4}H\right) \frac{\partial^2 \check{x}}{\partial y^2} + (x_0 - x^*)\frac{\partial^2 \check{x}^2}{\partial y^2} + \frac{1}{3}\frac{\partial^2 \check{x}^3}{\partial y^2} - \frac{1}{4}\frac{\partial^4 \check{x}}{\partial y^4}\right] , \tag{8.7}$$

where $\Delta x_0 = (x_0 - x_{S1})(x_0 - x_{S2})$, $H = h/\varepsilon_0$.

Let us carry out a qualitative analysis neglecting all the terms on the right hand side except for the first one. In the absence of an external stirring field, when the system is on average beyond the spinodal, a positive effective diffusion coefficient with gradual equalization of fluctuations corresponds to the state

$$x_0 < x_{S1} \quad \text{or} \quad x_0 > x_{S2} .$$

If the average concentration is inside the spinodal region, i.e.,

8.2 Extending the Region of Absolute Instability

$$x_{S1} < x_0 < x_{S2},$$

the effective diffusion coefficient is negative, the monophase state is unstable and fluctuations grow. Thus, at

$$x_0 = x_{S1} \quad \text{or} \quad x_0 = x_{S2},$$

the diffusion coefficient reverses its sign. Correspondingly, x_{S1}, x_{S2} become boundaries of the spinodal region. The presence of an external action sets other boundaries for the sign change of the diffusion coefficient, corresponding to the new concentrations $\tilde{x}_{S1}$, $\tilde{x}_{S2}$. Obviously, these concentrations must satisfy the condition

$$(\tilde{x}_{S1} - x_{S1})(\tilde{x}_{S2} - x_{S2}) - \frac{1}{4}H = 0. \tag{8.8}$$

In the experimental study [104], it was stated that stirring does not change the critical concentration value. Hence,

$$\tilde{x}^* \equiv \frac{1}{2}(\tilde{x}_{S1} + \tilde{x}_{S2}) = x^*.$$

Taking this into account, from (8.8), we get concentrations corresponding to the new instability boundary effective spinodal:

$$\begin{aligned}\tilde{x}_{S1} &= x^* - \left[(x^* - x_{S1})^2 + \frac{H}{4}\right]^{1/2}, \\ \tilde{x}_{S2} &= x^* + \left[(x_{S2} - x^*)^2 + \frac{H}{4}\right]^{1/2}.\end{aligned} \tag{8.9}$$

It is clear from (8.9) that, with the growth of H, the difference $\Delta \tilde{x}_S = \tilde{x}_{S2} - \tilde{x}_{S1}$ increases. This means that the spinodal curve shifts towards the binodal, thereby extending the instability region of one-phase states (Fig. 8.2). On the other hand, it should be noted that, at some concentrations $\tilde{x}_1^*$ and $\tilde{x}_2^*$, the binodal and effective spinodal merge into a line. As a result the interval of metastable states vanishes in some finite vicinity of the critical concentration x^*. Thus, in the range of concentrations $\tilde{x}_1^*$ and $\tilde{x}_2^*$, the system can be pushed into the spinodal region by infinitesimal cooling, thus bypassing nucleus formation. The character of this range is such that it will be virtually impossible to distinguish the spinodal curve from the binodal. Using (8.6) and the definition of the binodal,

$$\frac{\partial f(x_B)}{\partial x} = 0,$$

we obtain for binodal concentrations

$$x_{B1} = x^* - \sqrt{3}(x^* - x_{S1}),$$

$$x_{B2} = x^* + \sqrt{3}(x_{S2} - x^*).$$

From the conditions

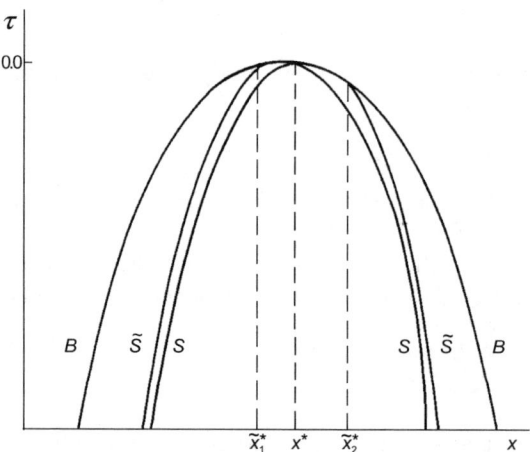

Fig. 8.2. Influence of external perturbation on extension of the spinodal region of a binary system: B binodal, S spinodal, $\tilde{S}$ spinodal arising under external perturbation ($H \approx 0.001$, $\Delta \tilde{x}_S \approx 0.03$)

$$\tilde{x}_1^* = \tilde{x}_{S1} = x_{B1}, \quad \tilde{x}_2^* = \tilde{x}_{S2} = x_{B2},$$

we find the required concentrations

$$\tilde{x}_1^* = x^* - (H/8)^{1/2}, \quad \tilde{x}_2^* = x^* + (H/8)^{1/2}. \tag{8.10}$$

In the case when the molecule concentration in a liquid $n \approx 1\,027$ m^{-3}, the intermolecular radius of the interaction is $b \approx 10^{-9}$ m. Hence, for $\varepsilon_0 = nk_\mathrm{B}T$ at $T \approx 300$ K, the estimate $\varepsilon_0 \approx 4.106$ J/m is valid. Making use of the expression (8.4) for the jump in the heat capacity,

$$\Delta C = \frac{a^2 \varepsilon_0^{1/2}}{8\pi T_c^2 K^{1/2} H^{1/2}}, \tag{8.11}$$

we evaluate H at $\omega = 1$ Hz and $\Delta C = 132$ J/kg K [104]. Supposing $K = \varepsilon_0 b$ and using (8.11) for H, we find

$$H = \left(\frac{a^2}{8\pi T_c^2 b \Delta C}\right)^2. \tag{8.12}$$

Thus, from (8.12) and taking into account (8.10), we can calculate the value $H \approx 0.001$. In this case the width of the merging region of binodal and spinodal lines (8.9) will be equal to 0.03 m.f., which is very close to the experimental value of 0.05 [104].

8.3 Initial Stage Kinetics of Forced Spinodal Decomposition

In the initial stages of spinodal decomposition, the solution composition at every point differs insignificantly from the average x_0. We can thus limit ourselves to the quadratic term in the expansion of $f(x)$ in $\check{x}$:

$$f(x) = f(x) + \frac{\partial f}{\partial x_0}\check{x} + \frac{\partial^2 f}{\partial x_0^2}\check{x}^2 .$$

Taking into account the condition

$$\int (x - x_0)\,\mathrm{d}V = 0 ,$$

the difference between the free energy F in the presence of concentration fluctuations and that of a homogeneous solution with composition x_0 at weak stirring is

$$\Delta F = F - f(x_0)V = \int \left[\frac{1}{2}\frac{\partial^2 f}{\partial x_0^2}\check{x}^2 + K(\nabla \check{x})^2 - h\check{x}^2\right] \mathrm{d}V .$$

In order to analyze the stability of the solution with respect to infinitesimally small changes in composition, it is convenient to make use of the Fourier representation. By expanding $\check{x}$ in the Fourier series

$$\check{x} = \sum_k A(k,t)\exp(ikr) ,$$

we obtain

$$\Delta F = \frac{1}{2}V\sum_k |A(k,t)|^2 \left(\frac{\partial^2 f}{\partial x_0^2} + 2Kk^2 - 2h\right) . \tag{8.13}$$

From the analysis of the solution stability with respect to infinitesimally small $\check{x}$, it follows from (8.13) that the solution is unstable with respect to concentration fluctuations with wavelengths

$$\lambda > \lambda_c = 2\pi \left[-\frac{1}{2K}\left(2h - \frac{\partial^2 f}{\partial x_0^2}\right)\right]^{-1/2} .$$

Let us consider the kinetics of spinodal decomposition in the early stages. We can write the non-linear Cahn–Hillard equation (7.2), taking into account the free energy functional (7.14), as follows:

$$\frac{\partial \check{x}}{\partial t} = L\left[\frac{\partial^2 f}{\partial x_0^2} - 2h - 2K\nabla^2 \check{x}\right]\nabla^2 \check{x} . \tag{8.14}$$

The solution of (8.14) is represented as a Fourier series expansion

$$\check{x} = \sum_k A(k,t)\exp(ikr) , \tag{8.15}$$

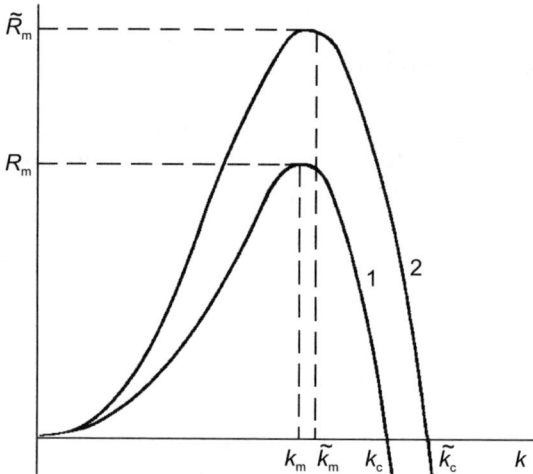

Fig. 8.3. Dependence of the Cahn amplification factor (curve 1) and the value calculated by (8.15) (curve 2) on the wave number

where $A(k,t) = A(k,0) \exp\left(\check{R}(k)t\right)$ and

$$\check{R}(k) = -Lk^2 \left[\frac{\partial^2 f}{\partial x_0^2} + 2Kk^2 - 2h\right] .$$

To elucidate the effect of the action h on spinodal decomposition, we compare $\check{R}(k,h)$ with the amplification factor in the absence of a stirring field (Fig. 8.3):

$$R(k) = -Lk^2 \left[\frac{\partial^2 f}{\partial x_0^2} + 2Kk^2\right] .$$

It is clear from (8.15) that the external action shifts the region of damping fluctuations ($k_c < k$) and the region of the fastest growing fluctuations ($k_m = k_c/\sqrt{2}$) towards larger values of k, corresponding to shorter wavelengths and smaller composition fluctuations. Values of the amplification factor increase linearly with the increase in H, and this gives rise to exponentially growing fluctuations. These fluctuations are also unstable and withdraw the system more rapidly from the metastable state. Slow modes, characteristic of the development of spinodal decomposition, fail to influence this process in the absence of an external field. Such a situation differs from the equilibrium case, where fluctuations vanish and the slowest ones prevail in the later stages. Thus, in the case of effective influence of an external field under non-equilibrium relaxation of the excited state, there appears the possibility of regulating the choice of 'surviving' fluctuations by the field, and hence also by the structure of the finite state, even in the initial stages of spinodal decomposition.

9. Microheterophase Relaxation State

9.1 Relaxation State
Near the Boundary of Absolute Instability
Under Weak Perturbation

A system undergoing phase transition under conditions of slow change of state is known to become macroscopically heterogeneous [302]. If this happens, the thermodynamic degrees of freedom fail to attain complete equilibrium. In this case, under a measured external action, one can obtain a metastable heterogeneous state consisting of several regions, each occupied by microscopically heterogeneous phases.

In Chap. 8 we considered the influence of weak stirring on local concentration fluctuations in a methanol–n-heptane mixture. It was established that, under stirring, the system has sections where the local concentration proves to be inside the spinodal, despite the fact that the system itself is, on average, beyond it. The size of the microscopically homogeneous sections depends on the stirring frequency. However, the problem of relaxation time in unstable states was not considered, whereas the system structurization depends on the rate of change of the state. In the case investigated, the diffusive-type dynamics of degrees of freedom may turn out to be a 'slow' process. This allows one to speak about separation of 'frozen' degrees of freedom from the background of fast-relaxing ones.

As a consequence, the whole system will go through a stage of slowly relaxing heterogeneous states. The scope of heterogeneity in the phase composition is set by the excited unstable state and the rate of change of system parameters. In [309] it is shown that such a state is not metastable and that changes in it can occur without the need to overcome barriers. It is thus concluded firstly that properties of the relaxation phase and microemulsion are to some extent similar. Secondly, complementary studies will explain how a microheterophase state is formed and how its parameters change. In the early stages of spinodal decomposition of the unstable state of a binary liquid mixture formed under weak perturbation.

Let us consider the continuity equation of a binary liquid with limited solubility:

9. Microheterophase Relaxation State

$$\frac{\partial x}{\partial t} + \mathrm{div} J = 0 , \tag{9.1}$$

where the flux is

$$J = -L\nabla\mu , \quad \text{with} \quad \mu = \frac{\partial F}{\partial x} , \tag{9.2}$$

and L is the kinetic Onsager coefficient characterizing molecular mobility.

The free energy functional of a binary solution under weak stirring is determined by (7.14), where

$$f = f_0 + \varphi \left[a(x - x^*)^2 + (x - x^*)^4 \right] . \tag{9.3}$$

The parameters a and x^* can be related to a concrete system as follows. The spinodal curve of a two-component solution is known to be approximated by the equation of state [14]

$$|x - x_c| = B|\tau|^\beta ,$$

where τ is the reduced temperature $(T - T_c)/T_c$. The asymmetry of the spinodal methanol–n-heptane system with respect to x_c [104] can be taken into account by a linear approximation. Then the equation of state of the spinodal concentrations $x_{S1,S2}$ will take the form

$$x_{S1,S2} = \pm B|\tau|^\beta + x_c - m|\tau| . \tag{9.4}$$

On the other hand, the following identity is valid for x_{S1} and x_{S2}:

$$\frac{\partial f(x_S)}{\partial x^2} = 0 ,$$

Using this with (9.3) and (9.4), we find that

$$a = -6B^2|\tau|^{2\beta} , \quad x^* = \frac{1}{2}(x_{S1} + x_{S2}) = x_c - m|\tau| . \tag{9.5}$$

Then using (9.1)–(9.3) in dimensionless variables $z' = cz/z_0$, $t' = t/t_0$, $H = h/\varphi$, $\chi = x - x'$ with characteristic scales

$$t_0 = \frac{z_0^2}{L\varphi} , \quad Z_0 = \left(\frac{K}{\varphi}\right)^{1/2} ,$$

and considering only the one-dimensional case, we get the continuity equation

$$\frac{\partial \chi}{\partial t} = 2\frac{\partial^2}{\partial z^2}\left[2\chi^3 - (H - \alpha)\chi - \frac{\partial^2 \chi}{\partial z_2} \right] . \tag{9.6}$$

In the early stages of decomposition, the fluctuations are small. Hence in (9.6), we can limit ourselves to the linear term with respect to χ. By introducing the fluctuation value $\xi = x(z,t) - x_0$, we arrive at the linearized diffusion equation

$$\frac{\partial \xi}{\partial t} = 2\frac{\partial^2}{\partial z^2}\left[(6\zeta^2 - H + \alpha)\xi - \frac{\partial^2 \xi}{\partial z^2} \right] . \tag{9.7}$$

9.1 Relaxation State Near the Boundary of Absolute Instability

where $\zeta = x_0 - x^*$. Representing the solution of (9.7) as a Fourier series,

$$\xi = \sum_k A(k,0) \exp(-ikr) \exp\left[-2k^2(6\zeta^2 - H + \alpha + k^2)t\right] , \tag{9.8}$$

we find that waves with wave number

$$k_m = \left(\frac{H - \alpha - 6\zeta^2}{2}\right)^{1/2}$$

grow the most rapidly.

Analysis of the solution (9.8) shows that, soon after the beginning of decomposition, one can neglect all the concentration waves except for those with wave number k_m. As a result, a characteristic scale of heterogeneities arises in the system, with wavelength

$$\lambda = \frac{2\pi}{k_m} = \left(\frac{8\pi^2}{H - \alpha - 6\zeta^2}\right)^{1/2} , \tag{9.9}$$

within which the concentration is locally homogeneous. Thus, at the linear stage of spinodal decomposition, the whole system is divided into small regions with locally homogeneous composition, and in the time interval $0 < t < t^*$, a cellular structure with effective extension λ is formed.

Let us evaluate the formation time of this structure. Qualitatively, this problem can be treated by using (9.6) to describe relaxation of the average value $\overline{\chi}$ of the variable χ over a cell. Supposing that

$$|\nabla \chi| \approx \overline{\chi}/\lambda , \qquad |\Delta \xi| \approx \overline{\chi}/\lambda^2 ,$$

and using (9.9), we find

$$\frac{\partial \overline{\chi}}{\partial t} = -\frac{2\overline{\chi}}{\lambda^2}\left(\lambda^{-2} + \alpha - H + 2\overline{\chi}^2\right) . \tag{9.10}$$

Substituting

$$\left|\frac{\partial \overline{\chi}}{\partial t}\right| \approx \frac{\overline{\chi}}{t^*} ,$$

in the linear approximation, we find the required time

$$t^* = \left(\frac{2\pi}{H - \alpha - 6\zeta^2}\right)^2 . \tag{9.11}$$

After this time has elapsed, the spinodal decomposition loses its linear character and the system goes into a non-linear relaxation stage. To consider the relaxation process itself, we integrate (9.10). Neglecting terms of second order, we have

$$\overline{\chi^2} = \frac{1}{2}\frac{H - \alpha}{1 - \exp\left[-\dfrac{(H - \alpha - 6\zeta^2)(H - \alpha)(t - t_0)}{2\pi^2}\right]} . \tag{9.12}$$

150   9. Microheterophase Relaxation State

From (9.12), for time $t \sim t^*$, x relaxes to one of the values

$$x = \pm \frac{1}{2}\left(H - \alpha - 6\zeta^2\right)^{1/2} + x_c - m|\tau| . \qquad (9.13)$$

From the results obtained in this way, the following conclusion can be made. The external perturbation H reduces the characteristic scale of the cellular structure in a binary liquid, as given by (9.9). Hence, local homogeneity and formation of the cells themselves are reached after a shorter time. This conclusion is also corroborated by the value obtained for the formation time of the cellular structure t^* [see (9.10)]. On the other hand, it follows from (9.8) and (9.13) that the perturbing field H causes a wider separation of cellular structures, because concentrations in cells relax to more disparate values, unlike the case when $H = 0$.

Using the parameters B and β determined in [184], equation (9.5) implies $\zeta = 0.02$, $H = 0.001$ (according to the evaluation performed in Sect. 8.2), $a = 6 \times 10^{-3}$, from which we find the formation time t^* and characteristic length λ of a cellular structure in the presence of a field. For the investigated system, the diffusion coefficient is $D \sim 10^{-7}$ and the characteristic length scale $z_0 \approx 10^{-8}$. Then the time scale is of order 10^{-9}–10^{-8} s. Substituting the above evaluation of the magnitudes into (9.9) and (9.11), we get

$$t^* \approx 0.01 \pm 0.1 \text{ s}, \quad \lambda \approx 10^{-6}\text{–}10^{-5} \text{ m} .$$

The relative change introduced by the field at $\omega = 1$ Hz into the values t^* and λ will be 10% of their absolute values.

9.2 Surface Tension Energy and Heat Absorption Effect

In the report [104], the system was assumed to transit from a spinodal regime to the nucleative regime for growth of heterogeneous microregions (drops), which is accompanied by the disappearance of pre-critical nuclei. The contribution of the drop surface tension to the system free energy must also change abruptly, and this may produce a marked jump in the heat capacity. Thus, the effect observed in [104] was regarded as being due to an endothermic change in the contribution of the surface tension energy to the total free energy of the system.

Let us prove the viability of this assumption using our model. For this purpose we investigate the frequency dependence of the surface energy Q of microheterogeneous regions. It was earlier supposed that a labile region, usually reached by a change in the temperature and revealing itself in the bulk, can also be obtained at fixed temperature under the action of an external stirring field, using the idea that the system is locally 'dragged' under the spinodal in small regions. Let us consider the situation when there is no external perturbation and a binary liquid system is in the equilibrium, separated state. Let the external stirring field start at some time $t = 0$. This

9.2 Surface Tension Energy and Heat Absorption Effect

will 'drag' the system into a one-phase state. As the field action increases, at certain instants of time, the degree of dispersion of the system will reach saturation for a given stirring frequency ω. At early times, the effective field H varies as

$$H(t) = \psi(t - t_0) \,.$$

At time t, cellular structures are formed, where

$$t^* \ll t - t_0 \,.$$

Therefore, the size of cells arising with increases in the field 'efficiency' is determined by the condition

$$\frac{H}{\Psi} \sim D^{-2} \,. \tag{9.14}$$

The diffusion coefficient for a linear stage in the process described by (9.7) is

$$D = 2(\alpha - H + 6\zeta^2) \,.$$

Substituting the latter into (9.14) for the length λ, we obtain

$$\lambda \sim \Psi^{-2/3} \,.$$

The quantity $\chi(z,t)$ is determined at every moment of time by the condition

$$\int dV = \chi_0 V = \text{Const.}$$

and by the requirement that the potential $F\{\chi\}$ be a minimum:

$$\Delta\chi = \left[2\chi^2 - [H(t) - \alpha]\right]\chi - g \,, \tag{9.15}$$

where g is the Lagrange multiplier for the constrained extremum problem. Equation (9.15) for $\chi(z)$ in the one-dimensional case can be represented as

$$\frac{\partial^2 \chi}{\partial z^2} = -\frac{\partial}{\partial \chi}\left[g\chi + \frac{1}{2}(H - \alpha)\chi^2 - \frac{1}{2}\chi^4\right] \,.$$

From the analogy with a known problem in mechanics [302], where $\chi(z)$ and z represent particle coordinates and time, we find the equivalent 'potential energy' $U(\chi)$ from

$$U(\chi) = g\chi + \frac{1}{2}(H - a)\chi^2 - \frac{1}{2}\chi^4 \,.$$

This has a minimum and two maximums. We are interested in solutions for which the turning points χ_1 and χ_2 are close to or coincide with maximums of $U(\chi)$. In this case the solution of (9.15) is an alternation of extended sections where $\chi \approx \chi_{1,2}$, and transitive regions with thickness

$$R \sim (H - \alpha)^{1/2} \,. \tag{9.16}$$

If the condition $R/\lambda \ll 1$ is satisfied for the characteristic length λ of the cellular structure, a transitive layer can be regarded as the phase-separation interface. At temperature τ and external field H, the magnitude of the surface tension σ satisfies

$$\sigma \sim \left[\frac{\chi_1 - \chi_2}{R}\right]^2 \sim (H - \alpha)^{3/2} .$$

Using the field frequency dependence $H = H_0 \omega^{-1/2}$ ($H_0 = 0.001$ for the stirring frequency $\omega = 1$ Hz) together with (9.9), we obtain for the surface energy

$$Q_S = \sigma S \sim \frac{(H_0 \omega^{-1/2} - \alpha)^{3/2}}{\lambda} , \qquad (9.17)$$

where $S(\lambda) \sim \lambda^{-1}$ is the total area of cells per unit volume.

For $\lambda \sim 10^{-6}$ m, $H = 0.001$, $\omega = 1$ Hz, the fractional change in the surface tension due to stirring is small. Although comparable with the jump in the heat capacity $\Delta C = 10\%$ [104], it remains less than the latter. However, the frequency dependence $Q_S(\omega)$ behaves in the same way as the dependence of the heat capacity jump. With increased stirring frequency, $\Delta C(\omega)$ and $Q_S(\omega)$ both go to saturation. Thus, the contribution of the surface energy is obviously an additional channel affecting the jump in the heat capacity.

9.3 Thermal Relaxation Effects in the Cellular Structure

In Chap. 7, the influence of an external perturbation on a system characterized by the singular behavior of the heat capacity was investigated in a narrow range of temperatures. Problems related to the influence of stirring at temperatures corresponding to a metastable state of the system were not considered. The analysis of experimental results [104] shows that the external action is also revealed in the mentioned region where the heat capacity depends on temperature. According to experimental data, there is an obvious tendency for the linear part of the heat capacity to shift towards larger values, depending on the intensity of the external perturbation. This fact can be explained on the basis of the model of the microheterogeneous relaxation state. At the beginning of Chap. 7, it was explained that, within the framework of the self-consistent field theory for a system with free energy (7.14), the mean squared concentration fluctuations of a solution take the form

$$\langle \delta x(r_1) \delta x(r_2) \rangle \sim \frac{1}{4\pi K r} \exp\left(-\frac{r}{r_c}\right) . \qquad (9.18)$$

The proportionality expressed in (9.18) points to the non-Poissonian character of local fluctuations caused by external stirring. Because of the correlation between nearby composition fluctuations, formation of a cellular structure is possible, over which the concentration will be uniform. Then the correlation radius r_c can play the role of the thickness R of a transitive layer separating cellular microregions with concentrations x_1 and x_2. The surface energy of cells depends on the external field as in (9.17):

$$Q \sim \frac{1}{\lambda}\left[\frac{x_1 - x_2}{R}\right]^2 \sim \frac{(H_0\omega^{-1/2} - \alpha)^{3/2}}{\lambda}. \tag{9.19}$$

In this region, the surface tension σ between simple liquids is changed since it is proportional to $\sigma_0 \sim k_B T/a^2$ (where a is the length of a molecule), and with the emergence of a heterophase state, σ is determined by the expression $\sigma_0 \sim k_B T/r_c^2$. This implies that if $r_c > a$, then $\sigma_0 > \sigma$. Hence, at times less than the time of relaxation to equilibrium, such a heterophase state must be similar to the microemulsion state. Relaxation of such a microemulsion to a homogeneous state in a closed system (adiabatic calorimeter) gives rise to an increase in temperature by $\Delta T = Q/C$. Thus, the formation of an unstable microemulsion on the phase-separation interface of two liquids can be revealed through measurements of the heat capacity. Using (9.19), we find that in our case ΔT depends on the external field via

$$\Delta T \sim \frac{(H_0\omega^{-1/2} - \alpha)^{3/2}}{\lambda C}.$$

As a result of such an exothermal effect, the heat capacity decreases and, at the end of the system relaxation, it is stabilized at a certain, lower value of C. This could appear in experimental data as a jump-like decrease in C at the temperature at which the mixer is turned off. With resumption of stirring, the microheterophase relaxation state must be restored. The heat capacity was measured by the adiabatic calorimetry method in the methanol–n-heptane system, for a separation state towards homogeneous, in the temperature range 300–325 K and at a relative concentration $x = 0.5$ m.f. of CH_3OH. The water content in the samples did not exceed $5 \times 10^{-5}\%$ and the measurement accuracy was not less than $\pm 0.5\%$.

The heat capacity was measured under stirring by an electromechanical mixer with changeable frequency, after reaching 309 K. The mixer was then turned off. The system was left to settle at this temperature for three hours and the heat capacity was measured in the temperature interval 309–313 K without stirring. At 313 K, the mixer was turned on again and, from this temperature up to 325 K, the measurements were carried out under stirring.

At 309 K the heat capacity decreased by about 5% (Fig. 9.1) and a monotonic change in C occurred at this level of values with increasing temperature. With the mixer on, a reverse process occurred at 313 K, with C increasing to values obtained under continuous stirring. One can evaluate the surface tension σ from the measurement data on C, considering that

$$\sigma \sim k_B T/r_c^2,$$

where r_c is the characteristic size of a microheterophase state. According to the measurement results, the surface energy Q of the heterophase state is

$$Q = \Delta T C.$$

9. Microheterophase Relaxation State

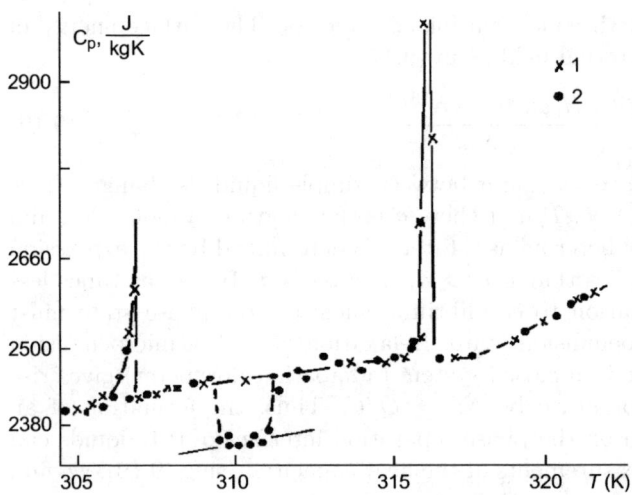

Fig. 9.1. Temperature dependence of the heat capacity of the methanol–n-heptane system: (1) under continuous stirring, (2) with the mixer off

On the other hand, $Q \approx n\sigma S$, where $S \sim r_c^2$ is the area of one cell, $n \approx V_m/r_c^2$ is the number of cells formed, V_m is the volume of a cellular phase, equal to 5–10% of the total system volume, $V_0 \sim 10^{-5}$ m^3, and $r_c \sim 10^{-6}$ m [104]. Then for ΔT,

$$\frac{\sigma}{\sigma_0} \approx \frac{r_c \Delta C}{V_0 \sigma_0} \geq 10^{-2}\text{--}10^{-3} \ .$$

Thus, as a result of weak mechanical stirring, a relaxation microemulsion state with low surface tension arises on the phase-separation interface of weakly soluble liquids.

10. Transition from Emulsion to Microemulsion

10.1 Microemulsion Structure

One of the main peculiarities of micelles is their ability to solubilize chemical compounds insoluble in a pure solvent. This extremely important practical property has given micelles a key role in technical activities. A nominal amount of solubilized substance proves to be enough to obtain highly swollen micelles, so that at some point they can be considered as small drops of an oil phase within a surfactant membrane in water. Under certain conditions, one can consider these solutions as two-phase systems and relate them to emulsions.

In these systems [162] the surface tension σ is a factor conditioning the stability and size of particles. An increase in the surface free energy σS under the appearance of a new phase is thermodynamically disadvantageous, and this does not therefore favor the formation of an emulsion. Consequently, it is important to know how to decrease σ. In industrial applications emulsifiers, i.e., surfactants, cosurfactants and various salts, are used to create stable emulsions. They have a specific molecular structure with polar and non-polar groups and are well soluble in a disperse system. Emulsions are obtained if the surfactant and auxiliary additives form a surface with small enough positive free energy.

In the production of emulsions, two independent problems arise:

- formation of new drops,
- stabilization of those drops during formation.

All techniques for obtaining emulsions are reduced to division of liquids into small drops by external action. However, emulsions can also be obtained without external influence. In most cases, when a drop of a heavier liquid is placed on the surface of a lighter one, spontaneous emulsification begins from unstable motions near the separation interface. Thin flows of the heavier liquid are randomly torn off the surface, penetrate into the lighter liquid and form small drops in it.

Actually, as was established by Kvinke in 1888, non-equilibrium conditions arise at the point of contact between two liquids, and the surface tension decreases. Such conditions can appear as a result of chemical compounds forming at separate points or heterogeneous absorption of the solution molecules on the surface. Turbulence spreads rapidly from these points. Threads of one liquid penetrate into the other, where they disintegrate into drops which then move inside each phase. However, there is a region of drop sizes that are below the limit for the existence of the two-phase state in a solution. In this region, swollen drops are stable objects, i.e., do not disintegrate into separate phases. Such solutions are usually considered to be in thermodynamic equilibrium and related to microemulsions.

Microemulsions are dispersions of oil and water obtained by adding amphiphiles. They differ from typical emulsions by their small size ($\sim$ 100 Å instead of 1 μm) and absolute stability. Obviously, microemulsions are more structured than typical emulsions. If the system is structured as a result of dynamical reconstruction, its entropy decreases in consequence of the fact that the system must relax to the initial state. In this case, however, microemulsions are more stable than two-component systems. This was first demonstrated by J. Shulman and co-workers [304], who started fundamental research into such systems. To explain the spontaneous dispersion of oil and water, the notion of negative surface tension was introduced. They supposed that the latter could explain the free energy decrease on the phase-separation interface during formation of a new phase, whereby microemulsions would be stabilized.

Later E. Rukenshtein et al. [24] demonstrated that microemulsion formation can occur without negative surface tension. The increase in dispersion entropy of the system completely compensates a small, but positive change in the surface energy when the new phase arises, i.e., the positive contribution to the free energy of microemulsion formation is compensated by negative contributions from the emergence entropy of a dispersion system, and in particular by the free energy of formation of double layers. These effects, as well as the positive contribution to the free energy made by the repulsive forces of double layers, imply the existence of a finite drop radius at negative minimum ΔG_M of the free energy of microemulsion formation. They thus established that microemulsions are thermodynamically stable systems. In [55], an expression was obtained for the entropy contribution E_R occurring due to the formation of a drop with radius R, viz.,

$$E_R \sim \frac{k_\text{B}T}{4\pi R^2}.$$

For comparison, the surface tension between simple liquids is of the order of $k_\text{B}T/a^2$ (where a is the size of a molecule). This means that if R is 100 times larger than the size of a molecule, then E_R is 10^4 times less than the usual surface tension. Therefore, the sizes of microemulsion drops must equal those of molecules. According to the theories of microemulsion, the curvature of the interface between the phases is due to the pressure difference across it. This pressure difference arises because oil molecules loosen hydrocarbon tails, and water molecules attract hydrophilic heads [306]. Under these conditions, the emerging pressure gradient causes the interface to twist until this difference is completely balanced by the Laplace pressure of the double layer. Depending on the type of stabilized drops, either water–oil (W/O) or oil–water (O/W), emulsions are obtained. An emulsifier can be characterized by a specific number, the hydrophile–lipophile balance (HLB). If the HLB is within the range 8–13, emulsions are O/W. Further increase in the content of surfactant does not affect the stability and size of emulsions.

The need for microemulsions with specific parameters is growing as wider applications are developed. Additional components, such as salts, cosurfactants and others, if introduced into the solution, can help to control the properties of disperse systems. According to the classification of all possible states of microemulsions constructed by P. Windsor [3], two-phase solutions are called systems of type I for a microemulsion in equilibrium with the organic phase and type II for a microemulsion in equilibrium with the aqueous phase. Three-phase solutions, in which microemulsions are in equilibrium with both phases, are systems of type III. By adding salt to a water–oil–amphiphile system, one can go over from one system to another.

Interest in disperse systems with adjustable properties has grown considerably due to the use of microemulsions in oil production. As explained in [307], the primary and secondary stages of oil production yield about 30% of the total amount. The remaining 70% must be extracted by tertiary production methods. The most promising is extraction using surfactant solutions. When a microemulsion contacts an oil phase, diffusion occurs. As a result of the ultra-low surface tension, oil merges with the microemulsion and becomes movable, forming traveling oil zones. However, it is difficult to achieve stability in such a system, so a small change in the brine concentration (a typical phenomenon in oil layers) causes liquid crystals to form. These are sometimes observed between the microemulsion and oil phases. As a result, mass transfer slows down and this impedes oil extraction. By increasing the amount of liquid components and changing external conditions, one can control properties of the system. Its stability, however, is difficult to attain.

It is known from the theory of systems that, with an increase in the number of interacting variables, the structural complexity of the system grows. This in turn can destroy its overall stability. The conflict between the stability and complexity of the system is regarded as resolved in favor of the former only if structural hierarchical division emerges within it.

Let us try to analyze the situation with multicomponent liquid solutions by taking into account the above discussion. In the investigated case, a transition is observed from a simple two-phase system, an emulsion containing two or three components, to a more complex one, a microemulsion consisting of five components or more. This transition is obviously accompanied by an increase in stability. We must therefore ask what hierarchical division is involved, and why this division in media with an easily changeable internal parameter can favor the stability of the system as a whole? In order to answer these questions, it seems that we must investigate relaxation processes in hierarchical systems, since the separation of mixed liquid solutions is directly determined by the kinetics and relaxation of chemical and molecular bond excitations [109].

Let us consider relaxation processes in a simple system. Its state is usually characterized by one parameter whose value at a given time depends on the system state at previous times. Since in a simple system there is no hierarchy of spatial distribution amongst the centers taking part in redistribution of the excitation, the excitation relaxation time has exponential dependence. As the system complexity increases, e.g., with a growing number of interacting components in a liquid solution, the kinetics of relaxation processes cannot be described by one parameter. Since the multicomponent system leads to the appearance of various relaxation times, in fact, to polychronal kinetics, their whole spectrum is available, from the fastest to the slowest. Obviously, fast relaxation times can bring a system autonomously to a quasi-equilibrium local state that will actually cause it to separate into distinct subsystems. However, it is also possible that the system excitation relaxation will not develop consecutively, but rather in a parallel way, provided that the spectrum of relaxation times is narrow enough. It may be that the kinetics of the system excitation relaxations in the first and second cases is different.

10.2 Polychronal Relaxation Processes and Dispersion on the Interface

The analysis of relaxation processes with a large number of parallel channels is described in a number of works, in which properties of disordered complex systems are simulated by fractal properties [308]. In particular, the transfer of excitations from a donor in condensed matter is investigated in [182]. The excitation attenuation law at a discrete donor caused by direct transfer of energy to a defect located at point R_i is considered. The relaxation function $\Phi_i(t)$, i.e., the probability that after time t a donor will remain in the excited state, depends on the rate of excitation relaxation $W(R_i)$. It turns out that for different hierarchies, excitation transfer rates and dependence $W(R_i)$, the function of excitation relaxation varies from a 'stretched' exponential law to an 'accelerated' power law. In any case, however, the pure exponential dependence characteristic of a disordered state, without hierarchical distribution of donors, was not obtained. Hence, the excitation attenuates non-exponentially because of the hierarchy of distances in the investigated system.

Such behavior is also observed when there are fast relaxation channels and the excitation transfer occurs towards adjoining defects. An extended exponential law is once again observed, only with much smaller exponent than in the previous case. The consecutive relaxation model, rather than the parallel relaxation model [1], is of special interest. This purely hierarchical model allows several stages of relaxation, with the faster degree of freedom

10.2 Polychronal Relaxation Processes and Dispersion on the Interface

relaxing before a slower one can begin. This means that the scale of the relaxation time at any chosen level is subject to relaxation at a lower level. Hence, the excitation relaxation in such a system is described by the slowest algebraic law, and the appearance of the hierarchical division in the system markedly slows down the rate of excitation relaxation.

Such a conclusion is highly significant because it suggests that hierarchical division of complex systems can provide their stability by virtue of their long relaxation times. It is easy to assume that, if a new state of a complex system forms as a result of its leaving an unstable state, then the appearance of infinitely long relaxation times due to hierarchical division can lead to a metastable state.

Let us examine this assumption in the case of a simple system, viz., the phase-separation boundary of two insoluble liquids. The emergence of the microheterophase state in a system containing immiscible components of various liquids is usually related to the addition of a surfactant [306]. The main feature of such a state is an internal microstructure determining a low surface tension. On the other hand, it is known [309] that if a system nevertheless transits into an unstable state, then it can, as a result of decomposition, end up in a metastable, slowly relaxing state with an internal microstructure. In this case one can expect a mixed state of insoluble liquids to appear. It will be shown below that such a state is brought about by relaxation of the finely dispersed state generated by mechanical stirring on the sharp boundary of phase-separated immiscible liquids.

While stirring, a system of two liquids with fine, dispersed particles is formed at the phase-separation interface. After turning off the perturbation, this system relaxes to the state of two liquids with a common boundary. In this connection, it is obvious enough that, under stirring, the common phase-separation boundary abruptly increases on account of fractal branching. Therefore, relaxation to some state with less boundary can be interpreted as a decrease in the interface energy. However, it is important to take into account the entropy contribution, which increases with the growth of system dispersion [82]. Therefore, the relaxation of the system must occur as a result of a decrease in the free energy:

$$F(\sigma) = E(\sigma) - TS(\sigma) ,$$

where E is the surface energy of the system and S is the entropy. Obviously, if δ is the average particle size, and $n \ll \infty$ the number of particles, then

$$E(\sigma) = \Omega n \sigma_{12} \delta^2 ,$$

where Ω is a dimensionless coefficient depending on the geometry of a particle (for a spherical shape $\Omega = 4\pi$), and σ_{12} is the surface tension between two liquids. Calculation of the entropy is more complex and it can be evaluated as described below. If a liquid has N molecules, then

$$S = n \ln\left[\frac{n}{n+N}\right] + N \ln\left[\frac{N}{n+N}\right] .$$

10. Transition from Emulsion to Microemulsion

Then, since

$$\delta^3 \approx \frac{1}{n} \quad \text{and} \quad N \gg n,$$

we get

$$F(\sigma) = \Omega \sigma_{12} \frac{1}{\delta} - bk_B T \frac{1}{\delta^3},$$

where b is a number bounded by 1. On the other hand,

$$\frac{dE}{dt} = \Omega \sigma_{12} \frac{d}{dt}\left(\frac{1}{\delta}\right) = -\Omega \sigma_{12} \frac{1}{\delta^2} \frac{d\delta}{dt}.$$

Thus,

$$\frac{d\delta}{dt} = J\left(1 - \frac{\xi}{\delta^2}\right) \tau_c,$$

where

$$\xi = 3bk_B \frac{T}{\Omega \sigma_{12}}$$

and J is a constant with units of $[\text{s}^{-1}]$. Let us consider particular cases.

- Regime A. Let $1 \gg \xi/\delta^2$, i.e., $\delta \gg \xi^{1/2}$. Then $\delta = \delta_0 + Jt$, i.e., the dispersion drops linearly (Fig. 10.1, curve 1).
- Regime B. Another marginal situation is

$$\frac{d\delta}{dt} = -J\xi \frac{1}{\delta^2},$$

which gives

$$\delta = \delta_0 \left(1 - 3J\xi \frac{t}{\delta_0^2}\right)^{1/3}.$$

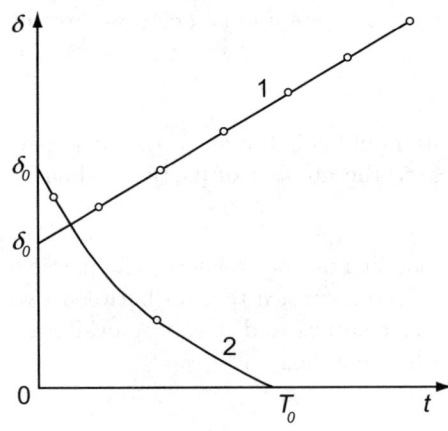

Fig. 10.1. Time dependence of the system dispersion

The characteristic 'dissolving' time is then

$$t_\mathrm{d} = \frac{\delta_0^3}{3J\xi}.$$

One can also obtain a general solution:

$$\frac{\delta^2 \mathrm{d}\delta}{\delta^2 - \xi} = Jt.$$

Then,

$$\delta + \xi \int \frac{\mathrm{d}\delta}{\delta^2 - \xi} = J\mathrm{d}t + \mathrm{Const.}$$

The calculation requires the inequality $\delta \gg \xi^{1/2}$ to be satisfied. Obviously, at $\xi \to 0$, we are in regime A, whilst in the opposite case ($\delta \ll \xi^{1/2}$), we have

$$\delta - \delta_0 = Jt + \mathrm{Const.},$$

which gives rise to regime B (Fig. 10.1, curve 2). Under adiabatic conditions, the mixed soluble state of weakly soluble liquids appearing as dispersion increases should be accompanied by heat release as a result of relaxation of the surface [309]. In fact it cannot, but is revealed when measuring the temperature dependence of the heat capacity, as evidenced experimentally [104] with stirring of weakly dissolved methanol–n-heptane liquids.

10.3 Stable Microheterophase State on the Interface of Weakly Dissolved Liquids

As mentioned above, a new stable state arises if relaxation of the excited state is accompanied by an increase in the entropy contribution. This is quite possible if evolution of the system is in line with its hierarchical separation. Polychronal kinetics of relaxation times must proceed via structural separation of the system. The active electronic subsystem in micellization gives rise to the smallest times in the spectrum of relaxation times due to localization of electron excitations in fluctuations in the surfactant monomer density. Hence, the whole system proves to be structured by division into electronic and molecular subsystems. In this case, the hierarchy of relaxation times takes shape naturally. It provides large relaxation times corresponding to the observed sequence of processes. For instance, micellization does not begin until formation of pre-micellar assemblies is completed. Then swelling, and hence a two-phase state, is possible only in the formed micelles. The natural hierarchical division of the whole system spontaneously arises from a stable microheterophase state of the microemulsion type as conditioned by the potential of this system for spontaneous structurization.

In this respect, it is of paramount importance to single out the hierarchy of the system as a separate controlled parameter, and also to simulate its influence on the microheterophase state of microemulsion type on the interface of two immiscible liquids. One of the main problems of microemulsion theory is the problem of thermodynamic stability. The challenge is considered to be a difficult one because, to begin with, we are dealing with a multicomponent system, and secondly, microemulsions are structured media, whose structure depends on the solution composition [306]. Therefore, the problem of structurization and stability are solved separately.

Meanwhile, it is known from the theory of systems [1] that it is the development of structure and the increase in the entropy conditioned by it that globally strengthens the system stability. Despite the fact that the stability mechanism was already associated with a structurization feature, viz., the dispersion of emulsion surfactant, in early studies [304, 311], at the present time, microemulsion stability is explained only by the active role of surfactants in the presence of alcohols and salts.

The study [78] concentrated on the appearance of a microheterophase state when the system undergoes a transition from an unstable state under non-equilibrium conditions. As was shown, under certain conditions caused by the kinetics and hierarchy of the excitation relaxation times, the system can hold thermodynamically stable states with a boundary separating regions of different order parameter. Hence, if system stability via structurization is a general property of disordered systems, then one can expect the emergence of microheterophase, structured states containing stable, mixed pseudo-phases divided by a transitive region with a low interphase tension on the interface of immiscible liquids without participation of surfactants.

As the subject of investigation, we chose the interface of immiscible liquids withdrawn from a stable state by mechanical stirring. Let us consider the relaxation to equilibrium on the basis of the lattice model of a binary mixture. Within the framework of the Brownian walk model of quasi-particles, a numerical model was developed to simulate the interface under stirring of two phases by a two-dimensional space. This space was taken as a square with cyclical boundary conditions. A random distribution of differently colored lattice sites (A is white and B is black) correspond to the initial stage of the mixed non-equilibrium state. During the process of relaxation, the instant of interface formation was traced between two infinite clusters of different colors. For every boundary, the following characteristics were determined:

- the probability $P(z)$ of finding the site of the interface at a distance z, which is located along the axis of concentration changes,
- the interface width h (the number of included sites),
- its dimensionality Θ,
- the site dimension $d - \Theta$, where d is the space dimension,
- the interface coordinate z.

10.3 Stable Microheterophase State

Quantities h and z are determined as follows:

$$h = \frac{\int\limits^{\infty}(z - z_\text{f})^2 P(z)\,\text{d}z}{\int\limits^{\infty} P(z)\,\text{d}z}, \quad \text{where} \quad z_\text{f} = \frac{\int\limits_0^{\infty} zP(z)\,\text{d}z}{\int\limits_0^{\infty} P(z)\,\text{d}z}.$$

Two cases were considered. In the first case, all relaxation times of the system were assumed to have comparable values. In the movement of particles, the gravitational forces $G(j)$ and interphase interaction forces $\Psi(i,j)$ were taken into account:

$$E = E_0\sigma + G(-j)\Psi(i,j)\,.$$

Testing by simulating phase separation, the relaxed state of the system was obtained as a sharp boundary. This evidences the fact that the stability of the system resulted from temporal reduction of component dispersion from one phase into another, with boundary width $h = 1.3$ (Fig. 10.2a).

The second case allowed a variety of relaxation times in the system. As a consequence, the degrees of freedom with larger relaxation times behave as though frozen. Hence, local regions appear in the system, each occupied by microscopically homogeneous phases. In this case, thermodynamic degrees of freedom fail to reach complete equilibrium. The whole system passes through stages of relatively slowly relaxing states that stimulate its dispersion. Following this approach to the kinetics of behavior, one needs to introduce an entropy term responsible for the effects of splitting off and mutual dissolution of phases:

$$S_\text{k} = \frac{\Theta}{H(i,j)}\,, \quad E = \sigma E_0 + G(-j)\Psi(i,j) - k_\text{B}TS_\text{k}\,,$$

where S_k is the entropy term taking into account cluster formation, Θ is a scale coefficient taking into account the entropy contribution, and $H(i)$ is the

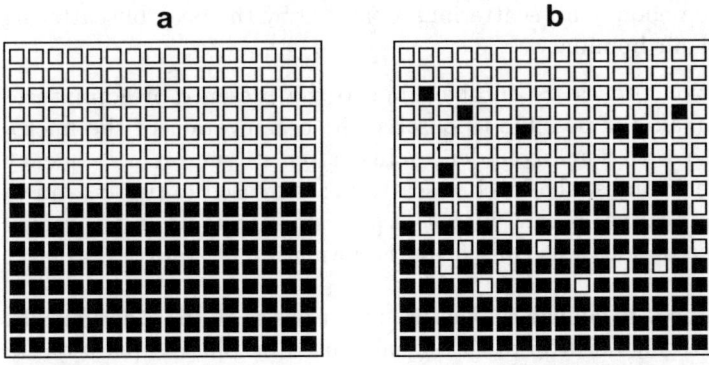

Fig. 10.2. Relaxed state of a system with comparable (**a**) and different (**b**) relaxation times

size of a formed cluster. A numerical model of cluster analysis was developed on the basis of the painting algorithm. Figure 10.2b shows a qualitatively new effect in the Brownian motion model, namely, the splitting off of clusters and formation of stable transitive state with boundary width $h = 2.11$ and characteristic fractal dimension 1.75. This state can be qualified as a foam-type microheterophase. In Fig. 10.2, one can see that apart from the microheterophase region, there are two others adjoining it, which can be defined as regions of mutual solubility of immiscible liquids.

The results of numerical simulations were compared with experimental data from direct optical observations of the microheterophase state arising on the interface of two immiscible liquids after preliminary mechanical stirring. The choice of experimental technique was thus based upon the simultaneous formation of a microheterophase layer, differing in its properties from the original liquids, and new boundaries. Hence, light-scattering observations seem to be justified as a method for registering this layer. In fact, its appearance must give rise to 'bifurcation' of the initial scattering peak on the interface. In addition, one can judge the altered interphase tension by the change in magnitude of the scattering signal I_p according to the known dependence $I_p \sim 1/\sigma$. For this purpose, periodic scanning was carried out in the direction perpendicular to the interface of the two liquids by the 'side-looking photometry' scheme, i.e., at an angle of 90° to the test signal direction. Laminar mechanical perturbation was carried out in the vertical direction across the interface at frequencies of 0.8–10 Hz, using an electromagnetic mixer that was removed from the dish during observations. Measurements were performed in a system of degassed water and filtered oil in the same volumes. These were consecutively poured into the dish with a quartz output window.

The scan of their initial state is shown in Fig. 10.3a (curve 1). Three regions are clearly distinguished:

- Region I: scattering by water,
- Region II: the peak of the scattering by oil, slightly widened due to unevenness of the boundary (the wetting meniscus),
- Region III: the region of air scattering, separated by the peak of scattering A from the oil–air boundary.

Here the relaxation process is durable, due to the presence of admixtures in the oil. This makes it possible to observe the organization of the microheterophase state with some confidence. Just after stirring at a frequency of 3 Hz for three minutes, the level of scattering increased abruptly, and in parallel with the peak A, a new, intense scattering maximum in the region of peak B appeared. This was caused by the formation of a stirred up layer of water–oil mixture.

Relaxation processes change the picture under observation rather quickly. In the course of a general drop in the signal caused by phase separation, in addition to the peak B, other new peaks appear. It is clear that sedimentation of water out of oil and sublimation of oil out of water both give rise to a

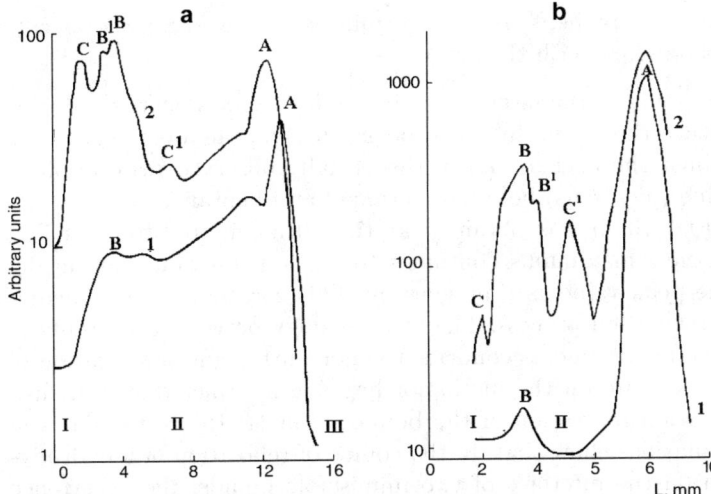

Fig. 10.3. Scan of scattering by degassed water–filtered oil (**a**) and methanol–heptane (**b**) boundary layers

natural decrease in the value of the scattered signal and to narrowing of the newly formed layer, or to an opposite shift of peak B and the new peaks. After stabilization of the process, in addition to possible emulsion formation, one should expect the emergence of transitive layers on both sides of the boundary (oil–water and water–oil) marked by their 'own' scattering maximums. One can also evaluate the capacity of these layers by the distance between them.

Curve 2 in Fig. 10.3a shows the scattering scan after stabilization of the relaxation process on the 14th day after stirring. The shift of peak A is caused by the inevitable removal of a small amount of oil when removing the mixer from the dish. Clearly, the pattern of scattering peaks fully corresponds to the accepted model: transitive phase I with peaks C–B', emulsion layer with peaks B'–B, and transitive phase II with peaks B–C'. Actually, the presence of a dip between peaks B and B' points to the formation of a layer with homogeneous optical properties, i.e., a constant refractive index, in the boundary region. Increase in the scattering signal evidences a significant drop in the interphase tension. The latter is also confirmed by visual observation of the meniscus at the visible water–oil boundary. Thus, curve 2 shows the stabilization of the microheterophase state due to simple mechanical stirring.

In the second experiment pure methanol–heptane liquids were investigated. In the scan of their initial state taken after settling (Fig. 10.3b, curve 1) two regions are clearly distinguished:

- Region I: scattering by methanol and by the methanol–heptane boundary layer, next to the scattering peak B widened by the unevenness of the interface,

- Region II: scattering by heptane, transforming to a powerful peak of scattering by the boundary with the air.

After the same stirring of the interface as in the first experiment, an abrupt growth of the scattering level and new peaks in the scan were registered. These occur against the background of the visually observed restoration of the separation line (meniscus) between methanol and heptane.

Curve 2 in Fig. 10.3b was obtained at the 80th minute after stirring. The intensity of new maximums continues to be non-constant, varying by up to 25% of the peak value as time goes by. This corroborates the incompleteness of the relaxation process. The total analogy between the results of the first and second experiments confirms the non-surfactant organization of a microheterophase state on the methanol–heptane interface due to hydrodynamic perturbation relaxation of the boundary layer. Hence the data on quantitative simulations show that, in the course of relaxation of a hydrodynamic excitation on the interface of two immiscible liquids, the occurrence of a metastable microheterophase state with low surface tension is actually possible as a result of structuring in the system in the absence of surfactants. In this case, a structured state represents a complex system divided by boundaries which separate the stable coexistence of pseudo-phases containing domains of various compositions and sizes.

10.4 Conclusion

Investigations of phase transitions and critical phenomena in binary phase-separating solutions and determination of thermophysical parameters in such systems can provide an adequate model for the description of more complex systems, e.g., multicomponent solutions with several phase transitions, micellar systems and microemulsions. The principal feature here is to explain the similarity of self-assembly processes and the stability of regular structures with phase transitions in terms of the universality of thermodynamic properties of objects close to second order phase transitions with the dominant correlation of abnormally growing fluctuations.

Moreover, studies of the non-linear dynamics of binary phase-separating solutions show that the appearance of a microheterophase state results from the introduction of such specific components as surfactants into the system, and also that it is a response to external perturbations. This may be decisive for further investigations into self-assembly of non-equilibrium processes in non-linear systems. Adjusting certain parameters of the external perturbation, an insignificant amount of which causes structuring, leads to either stabilization of the microemulsion state (introduction of surfactants, cosurfactants, etc.) or emergence of more complex dissipative and bicontinuous structures (e.g., with intensive mechanical stirring, hydrodynamic influences, laser irradiation, etc.).

Various methods of acoustic and optical spectroscopy, and adiabatic calorimetry applied to a wide range of liquid systems provided computations of the micellization thermodynamic parameters for solutions near critical points and merging phase transitions, and registered even insignificant changes in the solution structures. Further development of these methods and improved measurement accuracy will enable us to elucidate peculiarities of the phase behavior of multicomponent systems, and to prove some theoretical premises not yet confirmed experimentally.

However, processes occurring in multicomponent solutions are so diverse and conditioned by such a large number of degrees of freedom that they represent an immense area research for years to come. Although a unified theory of the liquid state is still under development, theoretical constructions describing liquid systems and processes of supramolecular organization advance studies of various types of system in the same way as methods traditionally used in other fields of science are successfully applied to the phenomena occurring in liquids.

References

1. J. Nikolis. *Dynamics of hierarchical systems. Evolution of representations*, p.488, [In Russian]. (Mir, Moskow 1989). H. Haken. *Synergetics*. 2nd ed, Springer Series in Synergetics, Vol.1. Springer, Berlin, New York, 1978.
2. M.A. Knackstedt and B.W. Ninham. *AICHE Journal* **41** (1995) 1295; M.A. Knackstedt and B.W. Ninham. Model disordered media provided by ternary microemulsions. *Phys.Rev.* **E 50** (1994) 2839–2843.
3. J. Bie, B. Klen, and P. Lalann. Phase diagrams and pseudophases. In: *Microemulsions. Structure, and Dynamics*, ed. by S.E. Friberg and P. Bothorel. Boca Raton, FL:CRC Press (1988).
4. D. Langevin. Low interface tension in microemulsions. In: *Microemulsions. Structure, and Dynamics*, ed. by S.E. Friberg and P. Botorel. Boca Raton, FL:CRC Press (1988).
5. J.M. Shulman and J.B. Montuge. *Ann.N.Y.Acad.Sci.* **92** (1961) 366.
6. J.E. Puig, E.I. Franses, and H.T. Davis. *Soc.Pet.Eng.J.* **19** (1979) 71.
7. P.D. Flemming, J.E. Vinatieri, and G.R. Glinsman. *J.Phys.Chem.* **84** (1980) 1526.
8. M. Kahlweit, E. Lessner, and G.R. Glinsman. *J.Phys.Chem.* **87** (1980) 1526.
9. M.W. Kim, J. Bock, and J.S. Huang. In: *Waves and Fluid Interfaces*, p. 151. Academic Press, New York (1983).
10. W.H. Wade, J.C. Morgan, and R.S. Schechter. *Soc.Pet.Eng.J.* **18** 1978) 242.
11. A.A. Saidov. Diagnostics of phase heterogeneities in liquid systems. Doctoral thesis (Tashkent 1991); P.K. Khabibullaev, A.A. Saidov, Sh.I. Mamatkulov, and Yu.V. Pakharukov. *Colloids and Surfaces A: Physicochemical and Engineering Aspects* **149** (1999) 242. V.S. Kononenko, S.Z. Mirzaev, A.A. Saidov, and P.K. Khabibullaev. *Acoustical Physics* **45** (1999) 108–110.
12. P. Honorat, D. Roux, and A.M. Bellocq. *J.Phys.Lett.* **45** (1984) 961.
13. N.F. Bunkin, O.A. Kiseleva, T.G. Movcha, B.W. Ninham, and O.I. Vinogradova. *Langmuir* **13** (1997) 3024; M.G. Neumann and G.L. de Sena. *Colloid Polym.Sci.* **275** (1997) 648–654; S. Senapati and A. Chandra. *J.Phys.Chem.* **B 105** (2001) 5106–5109.
14. M.A. Anisimov. *Critical Phenomena in Liquids and Liquid Crystals*. Gordon and Breach Publishers Science, 1991.
15. K. Wilson and J. Kogut. The Renormalization Group and the ϵ Expansion. *Phys.Rep.* **C 12** (1974) 75–199.
16. M.Kotlarchuk, Sow-Hsin Chen, and J.S. Huang. Critical behavior of a microemulsion studied by small-angle neutron scattering. *Phys.Rev.* **A 28** (1983) 508–511.
17. M. Corti, C. Minero, and V. Degiorgio. *J.Phys.Chem.* **88** (1984) 309.
18. R. Dorshow, F. Buzzaccarini, C.A. Bunton, and D.F. Nicoli. *J.Phys.Lett.* **47** (1981) 1336.

19. R. Zana and J. Lan. Dynamics of Microemulsions. In: *Microemulsions. Structure, and Dynamics*, ed. by S.E. Friberg and P. Botorel. Boca Raton, FL:CRC Press, 1988.
20. S. Atik and J.K. Thomas. *J.Phys.Chem.* **85** (1981) 3921.
21. P. Lianos, J. Lang, and A. Cazabat. In: *Surfactants in Solution*, ed. by P. Bothorel and K.L. Mittal, pp. 47. Plenum Press, New York, 1986.
22. A. Skoulios and D. Gullon. *J.Phys.Lett.* **38** (1977) 137.
23. S.A. Miller, R. Jvan, and W.H. Benton. *J. Colloid Interface Sci.* **56** (1976) 193.
24. E. Rukenstein. Stability, phase equilibrium and free energy of the interface surface of microemulsions. In: *Micellization, Solubilization, and Microemulsions*, ed. by K.L. Mittal. Plenum, New York, 1978.
25. J. Lang, A. Djavanbakht, R. Zana. In: *Microemulsions*, ed. by I.D. Robb. Plenum Press, New York, 1982.
26. E. Hirsh, F. Debeauvais, and F. Candau. *J. Physica* **45** (1984) 257.
27. I.I. Shinder. Equilibrium and non-equilibrium properties of double critical points of binary solutions in the presence of admixtures and external pressures. Abstract of PhD thesis. (Tashkent 1989).
28. P.P. Pugachevich and Yu.A. Khvorov. Density of heptane, undecane, hexadecane and their mixtures. *Otdel VINITI*, v. 3891-77.
29. M.M. Victorov. *Methods of calculation of physicochemical magnitudes and applied calculations*, p.284. Khimiya, Leningrad, 1977.
30. M.H. Karapetyanz. *Methods of comparative calculation of physicochemical properties*, p.528. Nauka, Moskow, 1965.
31. L.P. Filippov. *Methods of calculation and prediction of properties of substances*, 201. MGU, Moskow, 1988.
32. V.A. Kirillin and A.E. Sheindlin. *Investigation of thermodynamical properties of substances*, p. 368. Energiya, Moskow-Leningrad, 1963.
33. L.A. Davidovich et al. (*J.Phys.Chem.* **63** (1989) 359–363; S. Dutour, J.L. Daridon, and B. Lagourette. *International Journal of Thermophysics* **21** (2000) 174–184; A. Ritzl, L. Belkoura, and D. Woermann. *Phys.Chem.Chem.Phys.* **1** (1999) 53.
34. M.D. Tilicheva. Physicochemical properties of individual hydrocarbons. Gostoptechisdat, Moskow (1957); S. Nishikawa, N. Yokoo, and N. Kuramoto. *J.Phys.Chem.* **B 102** (1998) 4830–4834; H. Huang and S. Nishikawa. *J.Phys.Chem.* **A 104** (2000) 5910–5915.
35. I.G. Mikhailov, V.A. Solovyev, and Yu.N. Syrnikov. *Fundamentals of Molecular Acoustics*, p.516. Nauka, Moskow, 1964.
36. L.I. Mandelstam and M.A. Leontovich. *JETP* **7** (1937) 438.
37. K.F. Hertzfeld and T.A. Litovitz. *Absorption and Dispersion of Ultrasonic Waves*, p.486. Academic Press, New York, 1959; R. Abraham, M. Abdulkhadar, and C.V. Asokan. *J.Chem. Thermodynamics* **32** (2000) 1–16; S. Baluja and S. Oza. *Fluid Phase Equilibria* **178** (2001) 233–238.
38. W.P. Mason. *Physical Acoustics and its application in Science and the properties of Solids*. Princenton, D.Van Nostrand Company, Inc., Bell Laboratories Series, 1958; V. Gomis, F. Ruiz, and J.C. Asensi. *Fluid Phase Equilibria* **172** (2000) 245–259; H. Huang, S. Nishikawa. *J.Phys.Chem.* **A 104** (2000) 5910–5915
39. V.N. Kartzev. *J.Phys.Chem.* **50**, (1976) 764.
40. V.N. Kartzev, O.Ya. Samoylov, and V.A. Zabalin. *J.Phys.Chem.* **3** (1979) 757.
41. L.A. Davidovich, and I.I. Shinder. Sound propagation near a double critical point. In: *Materials of 13th International Congress*. Belgrade, August 1989.

42. L. Bergman. *Ultrasound and its application in science and technology*. Moscow, 1957.
43. D. Sette. *J.Chem.Phys.* **18** (1950) 1592.
44. L.D. Landau and E.M. Lifshitz. *Statistical Physics*, 3rd edn. p.583. Nauka, Moscow, 1976.
45. M.A. Anisimov, A.V. Voronel, and E. Gorodezki. *JETP* **60** (1971) 563.
46. A.Z. Patashinskii and V.L. Pokrovski. *Fluctuation theory of phase transitions*. Pergamon Press, Oxford UK, 1979.
47. R.G. Calflisch and J.S. Walker. Closed-loop phase diagrams, vaporization, and multicriticality in binary liquid mixtures. *Phys.Rev.* **B 28** (1983) 2535–2546; H. Doi, K. Tamura, and S. Murakami. *J.Chem. Thermodynamics* **32** (2000) 729–741.
48. J.C. Wheeler and G.R. Andersen. *J.Chem.Phys.* **73** (1980) 5778–5785.
49. J.C. Wheeler. *J.Chem.Phys.* **62** (1975) 433–439.
50. G.R. Andersen and J.C. Wheeler. *J.Chem.Phys.* **89** (1978) 3403–3413.
51. S.A. Vause and J.S. Walker. *Phys.Lett.* **A 90** (1982) 419–424.
52. Y. Harada, Y. Suzuki, and Y. Ishida. *J.Phys.Soc. Japan* **48** (1980) 705–706; X. An, X. Cui, W. Shen. *J.Chem. Thermodynamics* **32** (2000) 187–195; H. Katayama. *Fluid Phase Equilibria* **164** (1999) 83–95.
53. O. Bozdag and J.A. Lamb. *J.Chem. Thermodynamics* **15** (1983) 165–171; K. Rehak, J. Matous, J.P. Novak, and M. Bendova. *J.Chem. Thermodynamics* **32** (2000) 393–400; M. Sliwinska-Bartkowiak, S.L. Sowers, and K.E. Gubbins. *Langmuir* **13** (1997) 1182–1188.
54. D. Bernabe, A. Romero-Martinez, and A. Trejo. (Fluid Phase Equil. **40** (1988) 279–288; K. Rehak, J. Matous, J.P. Novak, and M. Bendova. *J.Chem. Thermodynamics* **32** (2000) 393-400.
55. P.K. Khabibullaev, V.S. Kononenko, S.Z. Mirzaev, and A.A. Saidov, I.I. Shinder. *DAN Uzbekistan* (1995) No. 2, 14–17.
    S.Z. Mirzaev, P.K. Khabibullaev, V.S. Kononenko, and A.A. Saidov. *J.Chem.Phys.* **112** (2000) 1057–1058; V.I. Ananchenkov, R.M. Galimzyanov, A.A. Saidov, and K.N. Kholov. *Colloids and Surfaces A: Physicochemical and Engineering Aspects* **149** (1999) 235–238; P.K. Khabibullaev, S.Z. Mirzaev, and A.A. Saidov. *Uzb.J.Phys.* **1–2** (1999) 40–46; V.S. Kononenko, S.Z. Mirzaev, A.A. Saidov, P.K. Khabibullaev, and I.I. Shinder. *Acoustical Physics Systems* **45** (1999) 108–110; S.Z. Mirzaev, P.K. Khabibullaev, A.A. Saidov, V.S. Kononenko, and I.I. Shinder. *J. Accoustic. Soc.Am.* **104** (1998), No. 7; A.A. Saidov. *Ultrasonic Studies of Micellar Systems. Encyclopedia of Surface and Colloid Science*. Marcel Dekker, New York, 2002.
56. A.G. Aizpiri, J.A. Correa, R.G. Rubio, and D.P. Mateo. Coexistence curve of methanol+n-heptane: Range of simple scaling and critical amplitudes. *Phys.Rev.* **B 41** (1990) 9003–9012.
57. R.B. Griffits. *J.Chem.Phys.* **60** (1974) 195.
58. J.C. Lang, Jr. and B. Widom. *Physica A* **81** (1975) 190.
59. L.A. Davidovich and I.I. Shinder. *JETP* **95** (1989) 1289–1301.
60. W. Koch, V. Dohm, and D. Stauffer. Order-Parameter Relaxation Times of Finite Three-Dimensional Ising-like Systems. *Phys.Rev.Lett.* **77** (1996) 1789–1792; F. Aliotta, M.E. Fontanella, M. Pieruccini, and C. Vasi. Aggregation phenomena in a lecithin-based gel: Transient networks and diffusional dynamics. *Phys.Rev.* **E 59** (1999) 665–672.
61. A. Kumar, S. Guha, E.S.R. Gopal. *Phys.Lett.* **A 123** (1987) 489–497.
62. S.V. Krivohija, I.L. Fabelinski, and L.L. Chaikov. *Izvestia vuzov.-ser. Radiophisika* **30** (1987) 308–316.
63. A.A. Sobyanin. *Uzbek.Phys.J.* **149** (1986) 325–328.

64. T.Kh. Ahmedov, L.A. Davidovich, I.I. Shinder, and P.K. Khabibullaev *Uzbek.Phys.J.* (1993) No. 5, 72–74.
65. I.I. Shinder, P.K. Khabibullaev, T.Kh. Ahmedov, and S.V. Tregulov. *Phys.Rev.* **49**, No. 6, (1994).
66. M.I. Shakhparonov, Yu.G.Shoroshev, S.S. Aliev, P.K. Khabibullaev, and L.V. Lanshina. Kinetics of Concentration Fluctuations in Isooctane-Nitroethane Solutions Having a Critical Stratification Point. *JETP Lett.* **7** (1968) 315–317.
67. P.K. Khabibullaev and S.S. Aliev *J.Phys.Chem.* **43** (1969) 2543–2548.
68. C. Garland. Ultrasound investigations of phase transitions and critical points. In: *Physical Acoustics*, ed. by P. Mason. Academic press, New York & London, 1966.
69. A.E. Brown and E.G. Richardson. *Philos.Mag.* **4** (1959) 705–720.
70. P.K. Khabibullaev and L.A. Davidovich *Acoust.J.* **18** (1972) 470–472.
71. L.A. Davidovich Investigation of ultra- and hyperacoustic properties of solutions with critical phase-separation point and of some organic liquids [In Russian]. Abstract of PhD thesis. Tashkent, 1974.
72. P.K. Khabibullaev and B. Izbasarov *Acoust.J.* **20** (1974) 130–132.
73. F. Eggers *Acustica* **19** (1967) 323–328.
74. F. Eggers and T. Funk. *Pribori dlya nauchnikh issledovaniy* **44** (1973) 38–47.
75. V.S. Kononenko and V.F. Yakovlev. *Ultrasvukovaya technika*, (1965) No. 1, 20–25.
76. V.S. Aggarwal and A.K. Gupta *J.Phys.D: Applied Physics* **8** (1975) 2232.
77. J.F. Steven, S.S. Yun *J.Acoust. Soc.Am.* **83** (1988) 1384–1387.
78. P.K. Khabibullaev, E.V. Gevorkian, A.S. Lagunov. *Rheology of Liquid Crystals*. Tashkent FAN, 1992; P.K. Khabibullaev. *Rheology of Liquid Crystals*. Alerton Press, New York, 1994.
79. A.E. Aliev, V.F. Krivorotov, and P.K. Khabibullaev *Fizika Tverdogo Tela (FTT)* **39** (1997) 1548.
80. M.A. Kasymdzhanov, S.S. Kasimova, S.S. Kurbanov, and P.K. Khabibullaev. *Ber. Bunsenges. Phys.Chem.* **101** (1997) 1.
81. P.J. Baymatov, P.K. Khabibullaev. *Fizika Tverdogo Tela (FTT)* **39** (1997) 441.
82. P.K. Khabbullaev et al. *DAN Russia* **354** (1997) No. 5.
83. I.I. Shinder, V.N. Khudaiberdiev, and L.A. Davidovich *Izvestiya AN USSR, Ser. phis.-mat. Nauk* (1982) No. 6, 74–75.
84. T. Kenshiro, N. Katsuo. *Appl. Phys.* **14** (1975) 149–150.
85. Yu.V. Valkov, M.F. Achilov, M.G. Khaliulin, and P.K. Khabibullaev. *Izvestiya AN UzSSR, Ser. phis.-mat. Nauk* (1973) No. 3, 98–100.
86. N. Prasad, S. Prakash, and K.S. Devivedi. *Acustica* **39** (1978) 276–279.
87. L.A. Davidovich, Kh. Rikhsitillaev, and P.K. Khabibullaev. Installation for investigation of hypersound velocity in liquids by Bragg refraction method. *Sbornik IX Vsesoyuznoy akusticheskoy konferentzii*. Moscow, part 2, 49–51, 1977.
88. V.A. Shutilov. *Fundamentals of Ultrasound Physics*, ppp. 191–193. LGU, Leningrad, 1980.
89. F. Eggers and Th. Funck. *Scientific Instruments* **44** (1973) 969.
90. T. Nishi and Y. Wada. *Jap.J.Appl.Phys.* **23** (1984) Suppl. Proc. 4 Symp.; *Ultrason. Electron.* Tokyo, 60–63, 1983.
91. A.R. Issam, S.S. Yun, and F.B. Stumpf. *J.Acoust.Soc.Amer.* **88** (1990) 1831–1836.
92. L.A. Davidovich, T.Kh. Ahmedov, I.I. Shinder, and M.K. Karabaev *Izvestiya AN UzSSR, Ser. phis.-mat. Nauk* (1986) No. 6, 56–60.

93. U. Kaatze, S. Trachimow, R. Pottel, and M. Bray. Broadband Study of the Scattering of Ultrasound by Polystyrene-latex-in-water suspensions. *Ann.Phys.(Leipzig)* **5** (1996) 13–33.
94. A. V. Kityk, W. Schranz, A. Fuith, D. Havlik, V.P. Sopronyuk, and H. Warhanek. Acoustic dispersion of $(NH_3C_2H_5)_2MnCl_4$ near the structural phase transition at 226 K. *Phys.Rev.* **B 53** (1996) 3055–3060.
95. V. Ilgunas, E. Yaronis, and V. Sukazkas. *Ultrasound Interferometers*, p.144. Vilnus [In Russian], Mokslas, 1983.
96. V.S. Kononenko. *Acoustic.J.* **33** (1987) 688–694.
97. F. Eggers, T. Funk, and L.H. Rikhman. *Pribory dlya nauchnikh issledovaniy* [In Russian] **47** (1976) 361–367.
98. V.S. Kononenko. *Uzbek.Phys.J.* (1992) No. 3, 44–48.
99. I.I. Shinder, V.N. Hudaiberdyev, and L.A. Davidovich. *Acoustic method of resonance reverberation in small volumes of liquids*. 25th Conference on Acoustics, Ultrasound'86. Bratislava **1** (1986) 98–103.
100. K. Kawasaki. Sound Attenuation and Dispersion near the Liquid-Gas Critical Point. *Phys.Rev.* **A 1**. (1970) 1750–1757.
101. R.A. Ferrell and J.K. Bhattacharjee. Dynamic scaling theory of the critical attenuation and dispersion of sound in a classical fluid: The binary liquid. *Phys.Rev.* **A 31** (1985) 1788–1809.
102. K. Kawasaki, Y. Shiwa. *Physica* **A 113** (1982) 27–43.
103. J.K. Bhattacharjee and R.A. Ferrell. Dynamic scaling theory for the critical ultrasonic attenuation in a binary liquid. *Phys.Rev.* **A 24** (1981) 1643–1647.
104. O.M. Atabaev, A.A. Saidov, P.T. Tadjibaev, Sh.O. Tursunov, and P.K. Khabibullaev. *DAN SSSR* **315** (1990) 889–891.
105. T.L. Hill. Thermodynamics of Small Systems. Vol.I. Benjamin, New York, 1963.
106. M.I. Leontovich. *Introduction to Thermodynamics. Statistical Physics*, p.426. Nauka, Moskow, 1986.
107. S.Z. Mirzaev. *Critical and noncritical low-frequency relaxational processes near liquid–liquid phase transitions*. Abstract of PhD thesis. Tashkent, 1996.
108. I. Procaccia and M. Gitterman. Dynamical critical phenomena in chemically reactive fluid mixtures. *Phys.Rev.* **A 25** (1982) 1137–1146.
109. I. Prigogine, P. Defay. *Chemical Thermodynamics*. Longmans Green, London,1954.
110. J.C. Wheeler and R.G. Petschek. Anomalies in chemical equilibria near critical points of dilute solutions. *Phys.Rev.* **A 28** (1983) 2442–2448.
111. V.A. Syrbu, M.I. Shahparonov. *Vestnik MGU. Ser. Khimiya* **3** (1973) 14, 299.
112. E. Illiel, N. Elinger, S. Enjial, and G. Morisson. *Conformational Analysis*. Mir, Moscow, 1969.
113. D.A. Rasulmuchammedova, P.K. Khabibullaev, and M.G. Haliulin. *Acoustic.J.* **22** (1976) 755–761.
114. E.N. Remsden. *The Beginning of Contemporary Chemistry*, p.784. Khimiya, Leningrad, 1989.
115. I.R. Krichevski, N.R. Khasanova, and L.N. Lifshitz. *DAN SSSR* **99** (1954) 113–115.
116. I.R. Krichevski and Yu.Yu. Zehanskaya. *J.Phys.Chem.* **30** (1956) 2315–2318.
117. P. Noyes. In: *Progress in Reaction Kinetics*. Pergamon Press, New York, 1961.
118. C. Reichardt. *Solvents and Solvent Effects in Organic Chemistry*. VCH, 1988.
119. G. Becker and F. Kohler. *Monatshefte für Chemie* **102** (1972) 556–570.
120. E. Koldin. *Fast Reactions in Solution*, p.310. Mir, Moskow, 1966.
121. I.L. Fabelinski. *Molecular Scattering of Light*, p.511. Nauka, Moskow, 1965.
122. I.L. Fabelinski. *UFN* **167** (1997) 721–733.

123. L.S. Ornstein and F. Zernike. *Phys. Zeitschrift BD* **19** (1918) 134–137.
124. M.F. Vuks. *Light Scattering in Gases, Liquids, and Solutions*, p.320. LGU, Leningrad, 1977.
125. N.Yu. Golubovski. *Investigation of integral intensity of molecular light scattering in binary liquid solutions in the vicinity of a critical phase-separation point* [In Russian]. Abstract of PhD thesis. Kiev, 1970.
126. R.G. Johnson, N.A. Clark. *Proc.Amer.Phys.Soc.* **54** (1985) 49–52.
127. V.P. Zaĭtsev, S.V. Krivokhizha, I.L. Fabelinskiĭ, A. Tsitrovskiĭ, L.L. Chaĭkov, E.V. Shvets, and P. Yani. Correlation Radius Near the Critical Points of a Guaiacolglycerin Solution. *JETP Lett.* **43** (1986) 85–88.
128. L.A. Davidovich, I.I. Shinder, and F.Kh. Abdullaev. *Method of measuring pressure and installation for its realization.* (Otkritiya, isobreteniya, No. 30, 159, A.S. 14833297, SSSR, 1989.
129. G.A. Larsen and S.M. Sorensen. Shear-Viscosity Behavior near the Double Critical Point of the Mixture 3-Methylpyridine, Water, and Heavy Water. *Phys.Rev.Lett.* **54** (1985) 343–345; P.A. Cirkel, J.P.M. van der Ploeg, and G.J.M. Koper. Branching and percolation in lecithin wormlike micelles studied by dielectric spectroscopy. *Phys.Rev.* **E 57** (1998) 6875–6883.
130. P. Salmettes. Critical Transport Properties of Fluids. *Phys.Rev.Lett.* **39** (1977) 1151–1154.
131. D.W. Oxtoby and W.M. Gelbart. *J.Chem.Phys.* **61** (1974) 2957–2963.
132. H. Tanaka and Y. Wada. Theoretical consideration on the acoustic anomaly of critical binary mixtures. *Phys.Rev.* **A 32** (1985) 512–524.
133. D.W. Oxtoby. *J.Chem.Phys.* **62** (1975) 1463–1468.
134. D.W. Oxtoby. *J.Chem.Phys.* **50** (1975) 459–460.
135. C.W. Garland and K. Nichigaku. *J.Chem.Phys.* **65** (1976) 5298–5301.
136. R.A. Ferrell and J. Bhattacharjee. Dynamic Scaling of Ultrasonic Attenuation at the Liquid Helium lambda Point. *Phys.Rev.Lett.* **44** (1980) 403–406.
137. M. Blandamer, N. Hidden, M.S. Symons, and N.S. Treloar. *Trans. Farady Soc.* **66** (1969) 3242–3246.
138. H. Haken. *Synergetics*, 3rd ed. Springer, Berlin, Heidelberg, 1983.
139. D.J. Mitchell and B.W. Ninham. *J.Chem.Soc. Faraday Transaction II* **77** (1981) 601.
140. M.I. Shahparonov. *Introduction to Contemporary Theory of Solutions*, pp. 56-57. Visschaya shkola, Moskow, 1976.
141. Z.N. Markina and L.P. Panycheva. *Kolloidniy Journal* **51** (1989) 696–699.
142. E. Ruckenstein and R. Nagarajan. *J.Phys.Chem.* **85** (1981) 3010–3014.
143. A.I. Rusanov. *Micellization in Surfactant Solutions*, p. 280. Khimiya, Sankt-Petersburg, 1992.
144. K.L. Mittel and P. Mukerdji. The wide world of micelles. In: *Micellization, Solubilization, and Microemulsions*, ed. by K.L. Mittal. New York, Plenum, 1978.
145. A.J. Frank. Radiative oxidation–reduction reactions in micellar solutions. In: *Micellization, Solubilization, and Microemulsions*, ed. by K.L. Mittal. New York, Plenum, 1978.
146. K. Kalianasundaram and J.K. Tomas. Radiation processes in non-ionic micelles. In: *Micellization, Solubilization, and Microemulsions*, ed. by K.L. Mittal. New York, Plenum, 1978.
147. M.A. Krivoglaz. *UFN* **111** (1973) 617–729.
148. I.A. Misurkin and A.A. Ovchinnikov. *JETP Lett.* No. 4 (1966) 248; A.D. MacKerell Jr. *J.Phys.Chem.* **99** (1995) 1846–1855.
149. V. Khaberdicil. *Substance Structure and Chemical Bond*, pp. 222–230. Mir, Moskow, 1974.

150. A.A. Ovchinnikov, I.I. Ukrainski, and G.G. Kventzel. *UFN* **108** (1972), 81–101.
151. F. Daniels and P. Alberti. *Physical Chemistry*, p. 648. Mir, Moskow, 1978.
152. L. Higgins, L. Salem. *Proc.Roy.Soc. (London)* **A 251** (1959) 172.
153. I.A. Misurkin, A.A. Ovchinnikov. *Jurnal strukturnoy chimii* No. 5 (1964) 888.
154. N.A. Popov. *Jurnal strukturnoy chimii* No. 10 (1969) 533.
155. Yu.A. Bychkov et al. *Jurnal experimentalnoy i teoreticheskoy khimii* **34** (1965) 739; R. Triolo, A. Triolo, F. Triolo, D.C. Steytler, C.A. Lewis, G.D. Wignall, and J.M. DeSimone. Structure of diblock copolymers in supercritical carbon dioxide and critical micellization pressure. *Phys.Rev.* **E 61** (2000) 4640–4643.
156. I.M. Lifshitz, S.A. Gradeskul, and L.A. Pastur. Introduction to the Theory of Disordered Systems. (Nauka, Moskow, 1982) 456.
157. M.A. Margulis. *Ultrasonics* **23** (1985) 157.
158. V.S. Letohov and V.P.Qebotarev. *Nelineaenay Lazernay Spektroskopiy Sverhvysokogo Rasrexeniy.* Nauka, Moskow, 1990.
159. Yu.M. Abrukina, B.L. Oksengendler. *Phisicheskiy Jurnal* No. 6 (1993) 53-58.
160. S. Fudzinaga. *Method of Molecular Orbitals*, p. 461. Mir, Moskow, 1983.
161. P. Mukerdji. Size distribution of micelles: Equilibrium of monomer micelles, processing of experimental data by molecular masses, sphere–cylinder transition and simple model of association. In: *Micellization, Solubilization, and Microemulsions*, ed. by K.L. Mittal. New York, Plenum, 1978.
162. A.I. Rusanov. *Kolloidniy Jurnal* **43** (1981), 903–911.
163. A.I. Rusanov. Surface tension of micelles and distribution by sizes. *Uspekhi kolloidnoi khimii*, pp. 139–145. ed. by B.S. Borisov. Kiev, 1983.
164. K. Shinoda. *Solvent Properties of Surfactant Solutions.* M.Dekker, New York, 1967.
165. A.A. Saidov et al. *Ingenerno-phisicheskiy jurnal*, No. 2 (1984) 59–61; A.P. Voleishis, L. Palchikova, A.A. Saidov, and P.K. Khabibullaev. *Ingenernophisicheskiy jurnal* **50** (1986) 76–85.
166. R. White, T. Gebell. *Far Order in Solids*, p. 447. Mir, Moskow, 1982.
167. H.E. Stanley. *Introduction to Phase Transitions and Critical Phenomena.* Clarendon Press, Oxford, UK, 1971.
168. J. Litster and R. Virgeno. Phases and Phase Transitions. *Phisika za rubejom* No. 3 (1983) 21–44.
169. P.K. Khabibullaev, Sh.I. Mamatkulov, Yu.V. Pacharukov, and A.A. Saidov. *DAN RF* **341** (1995) 756–758.
P.K. Khabibullaev, A.A. Saidov, Sh.I. Mamatkulov, and Yu.V. Pakharukov. Progress in Biomedical Optics. In: *Biomedical Sensing and Imaging Technologies* **3253** (1998) 282–285; M.G. Miguel, E. Marques, R. Dias, S.M. Melnikov, A. Khan, and B. Lindman. *Progr. Colloid Polym. Sci.* **112** (1999) 157–162; P.K. Khabibullaev, Sh.I. Mamatkulov, Yu.V. Pakharukov, and A.A. Saidov. *Colloids and Surfaces A: Physicochemical and Engineering Aspects* **149** (1999) 427–429.
170. Z.I. Markina and L.G. Isakovich . *Vestnik MGU* **30** (1989) 94–98 [In Russian]; F. Bockstahl, G. Duplatre. *J.Phys.Chem.* **B 105** (2001) 13–18.
171. R.E. Verral, D.J. Jobe , and E. Aicart. *J. Molecular Liquids* **65–66** (1995) 195–204.
172. A.E. Aripov, M.A. Orel, and S.N. Aminov. *Hydrophobic interactions in binary surfactant solutions*, p. 140, ed. by K.S. Ahmedov. FAN, Tashkent 1980.
173. Yu.A. Mirgorod. *Jurnal Phisicheskoi Khimii* **63** (1989) 195–204.
174. G.S. Hartley. *Aqueous Solutions of Paraffin-Chain Salts*, p. 136. Herman, Paris, 1936.
175. K. Shidegara. *Bull.Chem.Soc. Japan* **38** (1965) 1700.

176. E.A.G. Anianson and S.N. Wall. *J.Phys.Chem.* **78** (1974) 1224; M.G.Neumann and G.L. de Sena. *Colloid Polym.Sci.* **275** (1997) 648–654.
177. A.A. Saidov, V.I. Ananchenkov, and R.M. Galimzyanov *Uzbek.Phys.J.* No. 4 (1991) 648–654; V.I. Ananchenkov, R.M. Galimzyanov, A.A. Saidov, and K.N. Kholov. *Colloids and Surfaces A: Physicochemical and Engineering Aspects* **149** (1991) 235–238.
178. A.E. Aliev, A.A. Saidov, P.K. Khabibullaev, and I.I. Shinder *Akusticheskiy Jurnal* **42** (1997) 322–333; E.V. Chertkov, A.A. Saidov, P.K. Khabibullaev. In: *9th Intern. Conf. On Surface and Colloid Science*, Abstracts, p. 201, Sofia, Bulgaria, 1997; E.V. Chertkov, A.A. Saidov, and P.K. Khabibullaev. *Colloid and Interface Science A: Physicochemical and Engineering Aspects* **168** (2000) 185–191.
179. T. Matsuoka, T. Shibata, S. Koba, and H. Nomura. *J. Molecular Liquids* **65–66** (1995) 337–340; C. Baar, R. Buchner, and W. Kunz. *J.Phys.Chem.* **B 105** (2001) 2906–2913; T. Dalby and C.M. Care. Rosenbluth chain cluster growth in the study of micelle self-assembly. *Phys.Rev.* **E 59** (1999) 6152–6160; V.K. Aswal, S. De, P.S. Goyal, S. Bhattacharya, and R.K. Heenan. Small-angle neutron scattering study of micellar structures of dimeric surfactants. *Phys.Rev.* **E 57**(1998) 776–783.
180. A.P. Voleishis, V.N. Nishanov, M.K. Karabaev, A.A Saidov, and P.K. Khabibullaev. *DAN UzSSR*, No. 3 (1984) 27; A. Fogden, I. Carlsson, and J. Daicic. Beyond the harmonic bending theory of ionic surfactant interfaces. *Phys.Rev.* **E 57** (1998) 5694–5706; P. Tarazona, D. Duque, and E. Chacon. Aggregation models at high packing fraction. *Phys.Rev.* **E 62** (2000) 7147–7154.
181. P.K. Khabibullaev, Sh.I. Mamatkulov, Yu.V. Pacharukov, and A.A. Saidov. In: *9th Intern. Conf. On Surface and Colloid Science*, Abstracts, p. 373. Sofia, Bulgaria, 1997.
182. B. Lindman et al. *J.Phys.Chem.* **88** (1984) 5048; Y. Wang and R. Rajagopalan. *J.Chem.Phys.* **105** (1996) 696–705; A. Bhattacharya and S.D. Mahanti. *J.Phys.Condens. Matter* **13** (2001) 1413–1428; I. Borukhov and L. Leibler. *Phys.Rev.* **E** (1999) 41–44; I. Borukhov, R.F. Bruinsma, W.M. Gelbart, and A.J. Liu: *Phys.Rev.Lett.* (2000) 2182–2185; I. Borukhov. *Physica* **A 249** (1998) 315–320.
183. S.D. Baranovski, H. Friniz, E.I. Levin, B.I. Rusin, and B.S. Shklovski. *JETP* No. 4 (10) (1989) 2207.
184. J.S. Langer. *Ann.Phys.(N.Y.)* **65** (1971) 53–86.
185. V.P. Skripov and A.V. Skripov. *UFN* **128** (1979) 193–231.
186. Ya.B. Zeldovich and O.M. Todes. *JETP* **10** (1940) 1441–1445.
187. J.W. Cahn and J.E. Hilliard. *J.Chem.Phys.* **31** (1959) 688–699.
188. J.W. Cahn. *J.Chem.Phys.* **42** (1965) 93–99.
189. J.S. Langer, M. Bar-on, and H.D. Miller. New computational method in the theory of spinodal decomposition. *Phys.Rev.* **A 11** (1975) 1417–1429.
190. J.S. Langer. *Ann.Phys.(N.Y.)* **78**, (1973) 421–452.
191. J.S. Langer. Metastable States. *Physica* **73** (1974) 61–72.
192. J.S. Huang, W.I. Goldburg, and A.W. Bjerkaas. Study of Phase Separation in a Critical Binary Liquid Mixture: Spinodal Decomposition. *Phys.Rev.Lett.* **32** (1974) 921–923.
193. B.P. Skripov. Spinodal as envelope. In: *Phase transformations and non-equilibrium processes*, pp. 921–923. Sverdlovsk, 1980.
194. E.R. Smith and J.S. Rowlinson. *J.Chem.Soc.Faraday Trans. Part 2* **76** (1980) 921–923.
195. M.A. Novotny, W. Klein, and P.A. Rikvold. *Phys.Rev.* **B 33** (1986) 921–923.

196. K. Binder. *Physica* **A 140** (1986) 35–43.
197. N.F. Bunkin, F.V. Lobeev, and G.A. Lyahov *UFN* **167** (1997) 35–43.
198. G.F. Mazenko and O.T. Valls. *Phys.Rev.lett.* **59** (1987) 35–43.
199. A.I. Olemskoi, E.A. Toropov, and I.A. Sklyar *JETP* **100** (1991) 35–43.
200. J.C. Eilbeck, J.E. Further, and M. Grinfeld. *Phys.Lett.* **A 135** (1989) 35–43.
201. P.W. Bates, P.C. Fife. Spectral Comparison Principles for the Cahn–Hilliard and Phase-Field Equations, and the Time Scales for Coarsening. *Physica* **D 43** (1990) 335–348.
202. D.W. Heerman. *Phys.Rev.Lett.* **52** (1984) 35–43.
203. R. Toral, A. Chakrabarti, and J.D. Gunton. *Phys.Rev.Lett.* **60** (1988) 35–43
204. S. Puri. *Phase Transit.* **B 16–B 17** (1989) 35–43.
205. J.E. Farrel and O.T. Valls. Spinodal decomposition in a two-dimensional fluid model. *Phys.Rev.* **B 40** (1989) 7027-7039.
206. H. Furukawa. *Physica* **A 180** (1992) 128–155.
207. H. Jinnai, T. Koga, Y. Nishikawa, T. Hashimoto, and S. Hyde. *Proc.Amer.Phys.Soc.* **79** (1997) 128–155.
208. P. Fratze and O. Penrose. *Phys.Rev.* **B 40** (1989) 3477.
209. H. Ikeda. *Progr.Theor.Phys.* **61** (1979) 128–155.
210. O. Summich and H. Loffler. *Phys. Status Solidi.* **A 59** (1980) 128–155.
211. J.W. Hays and H. King. Fluctuation-induced spinodal decomposition in coherent metal-hydrogen systems. *Phys.Rev.* **B 25** (1982) 3298–3310.
212. A. Robledo and C. Varea. Kinetics of phase change in a model binary alloy. *Phys.Rev.* **B 25** (1982) 4711–4720.
213. Yu.N. Kurushin, Yu.P. Vorobyev, P.V. Geld, and A.N. Men. Analysis of phase dependence on composition taking into account structural and magnetic ordering. In: *Investigations of Metallurgical Processes*, pp. 128–155. Sverdlovsk, No.10, 1982.
214. V.S. Mitlin. *JETP* **95** (1989) 128–155.
215. C.A. Croxton. *J.Phys.C: Solid State Physics* **12** (1979) 128–155.
216. V.G. Baidakov. Stability of a metastable van der Vaals fluid. In: *Thermophysical Investigations of Overheated Liquids*, pp. 128–155. Sverdlovsk, 1981.
217. S.W. Koch, R.C. Desai, and F.F. Abraham. Spinodal decomposition of a one-component fluid: A hydrodynamic fluctuation theory and comparison with computer simulation. *Phys.Rev.* **A 26** (1982) 1015–1022; P. West, C. Trandum, and Y. Koga. *Biophysical Chemistry* **89** (2001) 53–63.
218. N.S. Andreev. *Physics and Chemistry of Glass* **4** (1978) 377–394.
219. R. Yokota: *J.Phys.Soc.Jap.* **45** (1978) 29–41.
220. C. Billotet and K. Binder. *J.Phys.B: Atomic and Molecular Physics* **32** (1979) 29–41.
221. C.C.J. Roothaan and J.H. Detrich. General quadratically convergent multi-configuration self-consistent-field theory in terms of reduced matrix elements. *Phys.Rev.* **A 27** (1983) 29–56.
222. F. Marsiglio and F.D. Manchester. *Phys.Lett.* **A 123** (1987) 70–81.
223. W. Klein. Fractals and multifractals in early-stage spinodal decomposition and continuous ordering. *Phys.Rev.Lett.* **65** (1990) 1462–1465.
224. J. Schemelzer. *Phys.Lett.* **A 158** (1991) 1462–1465.
225. E.D. Siggia. Late stages of spinodal decomposition in binary mixtures. *Phys.Rev.* **A 20** (1979) 595–605.
226. H.C. Burstyn, R.F. Chang, and J.V. Sengers. Nonexponential Decay of Critical Concentration Fluctuations in a Binary Liquid. *Phys.Rev.Lett.* **44** (1980) 410–413.
227. V.P. Koverda and G.N. Muratov. *Phase transformations and non-equilibrium processes.* Sverdlovsk, 1980.

228. R.A. Ferrell. Hydrodynamic singularity in the spinodal decomposition of a binary liquid. *Phys.Rev.* **A 24** (1981) 2814–2816.
229. R. Ruiz. Decay of concentration fluctuations in liquid mixtures. *Phys.Rev.* **A 27** (1983) 565–567.
230. G.I. Pojarskaya, N.L. Kasanova, Yu.D. Kolpakov, and V.P. Skripov. Phase-separation curve and spinodal of phase-separating n-heptane-perfluorhexane solution. In: *Thermophysical Properties of Metastable Systems.* Sverdlovsk, 1984.
231. A. Janosi. *J.Phys.B* **80** (1990) 393–400.
232. P.G. De Gennes. *J.Chem.Phys.* **72** (1980) 4756–4763.
233. P. Pincus. *J.Chem.Phys.* **75** (1981) 1996.
234. K. Hamono, M. Tachikawa, Y. Kenmochi, and N. Kuwahara. *Phys.Lett.* **A 90** (1982) 425–428.
235. B. Lindman and P. Stilbs. Molecular diffusion in microemulsions. In: *Microemulsions. Structure and Dynamics*, ed. by S.E. Friberg and P. Botorel. Boca Raton, FL:CRC, 1988.
236. V.S. Mitlin, L.I. Manevich, and I.Ya. Eruchimov. *JETP* **88** (1985) 495–506.
237. V.S. Mitlin and L.I. Manevich. *High-Molecular Compounds* **A 30** (1988) 9–15.
238. P. Wiltzius, F.S. Bates, and W.R. Heffner. Spinodal decomposition in isotopic polymer mixtures. *Phys.Rev.Lett.* **60** (1988) 1538–1541.
239. I.A. Nyrkova, A.P. Khohlov, and E.Yu. Kramarenko. *High-Molecular Compounds* **A 32** (1990) 918–926.
240. I.M. Abdulagatov, B.G. Alibekov. *J.Phys.Chem.* **57** (1983) 468–470.
241. V.F. Lysenkov [In Russian]. Issledovatelski phisicheski jurnal **48** (1985) 815–823.
242. V.G. Boyko, H.Y. Mogel, and A.V. Chaliy. *Ukrainian Phys.J.* **31** (1986) 137–143.
243. G.F. Mazenko, O.T. Valls, and F.C. Zhang. Renormalization-group theory of spinodal decomposition. *Phys.Rev.* **B 32** (1985) 5807–5817.
244. Y. Drossions and D. Ronis. Pseudospinodal critical phenomena, renormalized instantons, and the one-loop equation of state. *Phys.Rev.* **B 39** (1989) 12078–12097.
245. A.J. Bray. Exact renormalization-group results for domain-growth scaling in spinodal decomposition. *Phys.Rev.Lett.* **62** (1989) 2841–2844.
246. Ch. Roland and M. Grant. Monte Carlo renormalization-group study of spinodal decomposition: Scaling and growth. *Phys.Rev.* **B 39** (1989) 11971–11981.
247. Y. Yositaka and I. Akira. *J.Phys.Soc.Jap.* **45** (1978) 1949–1956.
248. D.S. Parmar and B. Labroo. *Phys.Lett.* **A 88** (1982) 1949–1956.
249. S.A. Jernov, A.F. Sirenko, and D. Hamana. FTT, Kiev-Donezk, No. 14 (1984).
250. A.F. Sirenko and D. Hamana. FTT, Kiev-Donezk, No. 15 (1985) 52–56.
251. I.I. Naumov. *Physics of Metals and Metallurgical Science* **60** (1985) 1139-1145.
252. M. Joshua, J.V. Maher, and W.I. Goldburg. Observation of Periodic Spinodal Decomposition. *Phys.Rev.Lett.* **51** (1983) 196.
253. M. Joshua and W.I. Goldburg. Periodic spinodal decomposition in a binary liquid mixture. *Phys.Rev.* **A 31** (1985) 3857–3867.
254. H.O. Carmesin, D.W. Heerman, and K. Binder. *J.Phys.B* **65** (1986) 89–102.
255. Y. Oono and S. Puri. Computationally efficient modeling of ordering of quenched phases. *Phys.Rev.Lett.* **58** (1987) 836–839.
256. J.E. Farrell and O.T. Valls. Spinodal decomposition in a two-dimensional fluid model: Heat, sound, and universality. *Phys.Rev.* **B 42** (1990) 2353–2362.
257. V.N. Semin. *J.Phys.Chem.* **63** (1989) 1102–1105.
258. V.N. Semin. *J.Phys.Chem.* **62** (1988) 2263–2266.

259. F.V. Bunkin, V.I. Podgaezki, and V.N. Semin. *Pisma v Jurnal Teoreticheskoi Phisiki* **14** (1988) 162-165.
260. M. Grant and J.D. Gunton. Temperature dependence of the dynamics of random interfaces. *Phys.Rev.* **B 28** (1983) 5496–5506.
261. S. Puri and Y. Oono. Effect of Noise on Spinodal Decomposition. *J.Phys.A: Math.Gen.* **21** (1988) L755–L762.
262. G.F. Mazenko, O.T. Valls, and M. Zannetti. Field theory of spinodal decomposition: Comparison with numerical simulations. *Phys.Rev.* **B 40** (1989) 379–383
263. R. Ruiz and D.R. Nelson. Anomalous Mixing Times in Turbulent Binary Mixtures at High Prandtl Number. *Phys.Rev.* **A 24** (1981) 2727.
264. A. Onuki and S. Takesue. *Phys.Lett.* **A 114** (1986) 133–136.
265. B.G. Emetz. *DAN USSR A* (1989) 56–57.
266. R. Ruiz and D.R. Nelson. Turbulence in binary fluid mixtures. *Phys.Rev.* **A 23** (1981) 3224–3246.
267. A. Onuki, A. Yamazaki, and K. Kawasaki. Light Scattering by Critical Fluids under Shear Flow. *Ann.Phys.(N.Y.)* **131** (1981) 217–242.
268. D. Beysens and M. Gbadamassi. Shear-Induced Effects on Critical Concentration Fluctuations. *Phys.Rev.* **A 22** (1980) 2250–2261.
269. A. Onuki. *Phys.Lett.* **A101** (1984) 286–290.
270. A. Onuki and K. Kawasaki. Critical Phenomena of Classical Fluids under Flow. I. *Progr.Theor.Phys.* **63** (1980) 122–139.
271. P. Guenoun, R. Gastoud, F. Perrot, and D. Beysens. Spinodal Decomposition Patterns in an Isodensity Critical Binary Fluid. Direct-Visualization and Light-Scattering Analyses. *Phys.Rev.* **A 36** (1987) 4876–4890.
272. C.K. Chan, W.I. Goldburg, and J.V. Maher. Light-scattering study of a turbulent critical binary mixture near the critical point. *Phys.Rev.* **A 35** (1987) 1756–1765.
273. S.V. Krivohija, O.A. Lugovaya, I.L. Fabelinski, and L.L. Chaikov. *JETP* **89** (1985) 85–91.
274. A. Onuki. *Physica* **A 140** (1986) 204–209.
275. C.K. Chan, F. Perrot, and D. Beysens. Effects of Hydrodynamics on Growth: Spinodal Decomposition under Uniform Shear Flow. *Phys.Rev.Lett.* **61** (1988) 412–415.
276. D.H. Rothman. Deformation, Growth, and Order in Sheared Spinodal Decomposition. *Phys.Rev.Lett.* **65** (1990) 3305–3308.
277. D. Beysens, M. Gbadamassi, and B. Moncef-Bouanz. New Developments in the Study of Binary Fluids under Shear Flow. *Phys.Rev.* **A 28** (1983) 2491–2509.
278. J.E. Denis and H.J.M. Hanley. *Phys.Rev.* **A 79** (1980) 178–180.
279. D. Beysens, R. Gastand, and F. Decruppe. Dichroism and birefringence induced by shear in a critical binary fluid. *Phys.Rev.* **A 30** (1984) 1145–1148.
280. D. Beysens and M. Gbadamassi. *Phys.Rev.Lett.* **47** (1981) 846–848.
281. Y.C. Chou and W.I. Goldburg. Birefringence from Periodic Shear Flow near the Critical Point. *Phys.Rev.Lett.* **47** (1981) 1155–1158.
282. A. Onuki. pinodal decomposition under shear. *Phys.Rev.* **A 34** (1986) 3528–3530.
283. Denis J.Evans and H.J.M. Hanley. A Thermodynamics of Steady Homogeneous Shear Flow. *Phys.Lett.* **A 80** (1980) 175–177.
284. T. Ohta, H. Nozaki, and M. Doi. *Phys.Lett.* **A 145** (1990) 304–308.
285. C.K. Chan and W.I. Goldburg. Late-Stage Pahse Separation and Hydrondynamic Flow in a Binary Liquid Mixture. *Phys.Rev.Lett.* **58** (1987) 674–677.

References

286. S. Starobinets, V. Yaknot, and L. Esterman. Critical dynamics of a binary fluid mixture in a centrifugal field. *Phys.Rev.* **A 20** (1979) 2582–2589.
287. D.S. Cannell. Measurement of the Long-range Correlation Length of SF_6 very Near the Critical Point. *Phys.Rev.* **A 12** (1975) 225–231.
288. S. Glen and D. Ronis. Critical phenomena in randomly stirred fluids: Correlation functions, equation of motion, and crossover behavior. *Phys.Rev.* **A 33** (1986) 3415–3432.
289. N. Easwar, J.V. Maher, D.J. Pine, and W.I. Goldburg. Active-Coupling Mixing Times for a Stirred Binary Liquid. *Phys.Rev.Lett.* **51** (1983) 1272–1274.
290. D.J. Pine, N. Faswar, J.V. Maher, and W.I. Goldburg. Turbulent suppression of spinodal decomposition. *Phys.Rev.* **A 29** (1984) 308–313.
291. J.A. Aronovitz and D.R. Nelson. Turbulence in phase-separating binary mixtures. *Phys.Rev.* **A 29** (1984) 2012–2016.
292. G. Satten and D. Ronis. Critical Phenomena in Randomly Stirred Fluids. *Phys.Rev.Lett.* **55** (1985) 91–94.
293. C.K. Chan, J.V. Maher, and W.I. Goldburg. Active-coupling mixing times for a stirred binary liquid. *Phys.Rev.* **A 32** (1985) 3117.
294. P. Tong, W.I. Goldburg, J. Stavans, and A. Onuki. *Phys.Rev.Lett.* **62** (1986) 91–94.
295. A. Onuki. On New Nonequilibrium Effects in Stirred Fluids. *Progr.Theor.Phys.Suppl.* **99** (1989) 382-398.
296. Yu.L. Klimontovich. *Statistical Physics*. New York, Harwood Academic Publ., 1986.
297. L.D. Landau and E.M. Lifshitz. *Course of Theoretical Physics*, Vol.6: Fluid Mechanics 2 Ed., Pergamon Press, New York, 1988.
298. E.P. Feldman and L.I. Stefanovich. *JETP* **96** (1989) 1513–1521.
299. P.K. Khabibullaev, M.Sh. Butabaev, Yu.V. Pacharukov, and A.A. Saidov. *DAN RF* **324** (1992) 1183–1186.
300. E.P. Feldman and L.I. Stefanovich. *JETP* **98** (1990) 1695–1704.
301. J.W. Cahn. *Trans.Metall.Soc. AIME* **242** (1968) 166.
302. A.Z. Patashinskii and I.S. Yakub. *FTT* **18** (1976) 3630–3636.
303. P.K. Khabibullaev, M.Sh. Butabaev, Yu.V. Pacharukov, and A.A Saidov. *DAN RF* **330** No. 2 (1995).
304. J.H. Schulman and E.G. Cockbain. *Trans. Faraday Soc.* (1940) 661.
305. P.K. Khabibullaev, Sh.I. Mamatkulov, Yu.V. Pacharukov, A.A. Saidov, and B.L. Oksengendler *DAN Uzbekistan* No. 3 (1995) 19–21.
306. S. Robben, L. Max. Theory of a phase state of microemulsions. In: *Micellization, Solubilization, and Microemulsions*, ed. by K.L. Mittel. New York, Plenum, 1978.
307. P. Neoji. Oil production and microemulsions. In: *Microemulsions. Structure and Dynamics*, ed. by S.E. Friberg and P. Bothorel. Boca Raton, FL:CRC Prss, 1988; L. Lobry, N. Micali, F. Mallamace, C. Liao, and Sow-Hsin Chen. nteraction and percolation in the L64 triblock copolymer micellar system. *Phys.Rev.* **E 60** (1999) 7076–7087; K.W. Jolley, M.H. Smith, N. Boden, and J.R. Henderson. *Phys.Rev.* **E 63** (051705, 2001); F. Aliotta, M.E. Fontanella, M. Pieruccini, C. Vasi. Aggregation phenomena in a lecithin-based gel: Transient networks and diffusional dynamics. *Phys.Rev.* **E 59** (1999) 665–672.
308. J.F. Gouye, M. Rosso, and B. Sapoval. Percolation in concentration gradient. In: *Fractals in Physics*, ed. by L. Pietronero and E. Tossatti. New York, Plenum Press, 1977.
309. P.K. Khabibullaev et al. *Uzbek.Phys.J.* No. 6 (1996) 36–39.
310. P.K. Khabibullaev, A.A. Saidov, et al. *Turkish J.Phys.* **21** (1997) 776.
311. A.A. Abramzon (Ed.). *Emulsions*, p. 520. Khimiya, Leningrad, 1972.

Springer Series in Solid-State Sciences

Editors: M. Cardona P. Fulde K. von Klitzing H.-J. Queisser

1 **Principles of Magnetic Resonance**
   3rd Edition By C. P. Slichter
2 **Introduction to Solid-State Theory**
   By O. Madelung
3 **Dynamical Scattering of X-Rays in Crystals** By Z. G. Pinsker
4 **Inelastic Electron Tunneling Spectroscopy**
   Editor: T. Wolfram
5 **Fundamentals of Crystal Growth I**
   Macroscopic Equilibrium and Transport Concepts
   By F. E. Rosenberger
6 **Magnetic Flux Structures in Superconductors**
   2nd Edition By R. P. Huebener
7 **Green's Functions in Quantum Physics**
   2nd Edition By E. N. Economou
8 **Solitons and Condensed Matter Physics**
   Editors: A. R. Bishop and T. Schneider
9 **Photoferroelectrics** By V. M. Fridkin
10 **Phonon Dispersion Relations in Insulators** By H. Bilz and W. Kress
11 **Electron Transport in Compound Semiconductors** By B. R. Nag
12 **The Physics of Elementary Excitations**
   By S. Nakajima, Y. Toyozawa, and R. Abe
13 **The Physics of Selenium and Tellurium**
   Editors: E. Gerlach and P. Grosse
14 **Magnetic Bubble Technology** 2nd Edition
   By A. H. Eschenfelder
15 **Modern Crystallography I**
   Fundamentals of Crystals
   Symmetry, and Methods of Structural Crystallography
   2nd Edition
   By B. K. Vainshtein
16 **Organic Molecular Crystals**
   Their Electronic States By E. A. Silinsh
17 **The Theory of Magnetism I**
   Statics and Dynamics
   By D. C. Mattis
18 **Relaxation of Elementary Excitations**
   Editors: R. Kubo and E. Hanamura
19 **Solitons** Mathematical Methods for Physicists
   By. G. Eilenberger
20 **Theory of Nonlinear Lattices**
   2nd Edition By M. Toda
21 **Modern Crystallography II**
   Structure of Crystals 2nd Edition
   By B. K. Vainshtein, V. L. Indenbom, and V. M. Fridkin
22 **Point Defects in Semiconductors I**
   Theoretical Aspects
   By M. Lannoo and J. Bourgoin
23 **Physics in One Dimension**
   Editors: J. Bernasconi and T. Schneider
24 **Physics in High Magnetics Fields**
   Editors: S. Chikazumi and N. Miura
25 **Fundamental Physics of Amorphous Semiconductors** Editor: F. Yonezawa
26 **Elastic Media with Microstructure I**
   One-Dimensional Models By I. A. Kunin
27 **Superconductivity of Transition Metals**
   Their Alloys and Compounds
   By S. V. Vonsovsky, Yu. A. Izyumov, and E. Z. Kurmaev
28 **The Structure and Properties of Matter**
   Editor: T. Matsubara
29 **Electron Correlation and Magnetism in Narrow-Band Systems** Editor: T. Moriya
30 **Statistical Physics I** Equilibrium Statistical Mechanics 2nd Edition
   By M. Toda, R. Kubo, N. Saito
31 **Statistical Physics II** Nonequilibrium Statistical Mechanics 2nd Edition
   By R. Kubo, M. Toda, N. Hashitsume
32 **Quantum Theory of Magnetism**
   2nd Edition By R. M. White
33 **Mixed Crystals** By A. I. Kitaigorodsky
34 **Phonons: Theory and Experiments I**
   Lattice Dynamics and Models of Interatomic Forces By P. Brüesch
35 **Point Defects in Semiconductors II**
   Experimental Aspects
   By J. Bourgoin and M. Lannoo
36 **Modern Crystallography III**
   Crystal Growth
   By A. A. Chernov
37 **Modern Chrystallography IV**
   Physical Properties of Crystals
   Editor: L. A. Shuvalov
38 **Physics of Intercalation Compounds**
   Editors: L. Pietronero and E. Tosatti
39 **Anderson Localization**
   Editors: Y. Nagaoka and H. Fukuyama
40 **Semiconductor Physics** An Introduction
   6th Edition By K. Seeger
41 **The LMTO Method**
   Muffin-Tin Orbitals and Electronic Structure
   By H. L. Skriver
42 **Crystal Optics with Spatial Dispersion, and Excitons** 2nd Edition
   By V. M. Agranovich and V. L. Ginzburg
43 **Structure Analysis of Point Defects in Solids**
   An Introduction to Multiple Magnetic Resonance Spectroscopy
   By J.-M. Spaeth, J. R. Niklas, and R. H. Bartram
44 **Elastic Media with Microstructure II**
   Three-Dimensional Models By I. A. Kunin
45 **Electronic Properties of Doped Semiconductors**
   By B. I. Shklovskii and A. L. Efros
46 **Topological Disorder in Condensed Matter**
   Editors: F. Yonezawa and T. Ninomiya

Springer Series in Solid-State Sciences
Editors: M. Cardona   P. Fulde   K. von Klitzing   H.-J. Queisser

47 **Statics and Dynamics of Nonlinear Systems**
   Editors: G. Benedek, H. Bilz, and R. Zeyher
48 **Magnetic Phase Transitions**
   Editors: M. Ausloos and R. J. Elliott
49 **Organic Molecular Aggregates**
   Electronic Excitation and Interaction Processes
   Editors: P. Reineker, H. Haken, and H. C. Wolf
50 **Multiple Diffraction of X-Rays in Crystals**
   By Shih-Lin Chang
51 **Phonon Scattering in Condensed Matter**
   Editors: W. Eisenmenger, K. Laßmann, and S. Döttinger
52 **Superconductivity in Magnetic and Exotic Materials**   Editors: T. Matsubara and A. Kotani
53 **Two-Dimensional Systems, Heterostructures, and Superlattices**
   Editors: G. Bauer, F. Kuchar, and H. Heinrich
54 **Magnetic Excitations and Fluctuations**
   Editors: S. W. Lovesey, U. Balucani, F. Borsa, and V. Tognetti
55 **The Theory of Magnetism II**   Thermodynamics and Statistical Mechanics   By D. C. Mattis
56 **Spin Fluctuations in Itinerant Electron Magnetism**   By T. Moriya
57 **Polycrystalline Semiconductors**
   Physical Properties and Applications
   Editor: G. Harbeke
58 **The Recursion Method and Its Applications**
   Editors: D. G. Pettifor and D. L. Weaire
59 **Dynamical Processes and Ordering on Solid Surfaces**   Editors: A. Yoshimori and M. Tsukada
60 **Excitonic Processes in Solids**
   By M. Ueta, H. Kanzaki, K. Kobayashi, Y. Toyozawa, and E. Hanamura
61 **Localization, Interaction, and Transport Phenomena**   Editors: B. Kramer, G. Bergmann, and Y. Bruynseraede
62 **Theory of Heavy Fermions and Valence Fluctuations**   Editors: T. Kasuya and T. Saso
63 **Electronic Properties of Polymers and Related Compounds**
   Editors: H. Kuzmany, M. Mehring, and S. Roth
64 **Symmetries in Physics**   Group Theory Applied to Physical Problems   2nd Edition
   By W. Ludwig and C. Falter
65 **Phonons: Theory and Experiments II**
   Experiments and Interpretation of Experimental Results   By P. Brüesch
66 **Phonons: Theory and Experiments III**
   Phenomena Related to Phonons
   By P. Brüesch
67 **Two-Dimensional Systems: Physics and New Devices**
   Editors: G. Bauer, F. Kuchar, and H. Heinrich

68 **Phonon Scattering in Condensed Matter V**
   Editors: A. C. Anderson and J. P. Wolfe
69 **Nonlinearity in Condensed Matter**
   Editors: A. R. Bishop, D. K. Campbell, P. Kumar, and S. E. Trullinger
70 **From Hamiltonians to Phase Diagrams**
   The Electronic and Statistical-Mechanical Theory of sp-Bonded Metals and Alloys   By J. Hafner
71 **High Magnetic Fields in Semiconductor Physics**
   Editor: G. Landwehr
72 **One-Dimensional Conductors**
   By S. Kagoshima, H. Nagasawa, and T. Sambongi
73 **Quantum Solid-State Physics**
   Editors: S. V. Vonsovsky and M. I. Katsnelson
74 **Quantum Monte Carlo Methods in Equilibrium and Nonequilibrium Systems**   Editor: M. Suzuki
75 **Electronic Structure and Optical Properties of Semiconductors**   2nd Edition
   By M. L. Cohen and J. R. Chelikowsky
76 **Electronic Properties of Conjugated Polymers**
   Editors: H. Kuzmany, M. Mehring, and S. Roth
77 **Fermi Surface Effects**
   Editors: J. Kondo and A. Yoshimori
78 **Group Theory and Its Applications in Physics**
   2nd Edition
   By T. Inui, Y. Tanabe, and Y. Onodera
79 **Elementary Excitations in Quantum Fluids**
   Editors: K. Ohbayashi and M. Watabe
80 **Monte Carlo Simulation in Statistical Physics**
   An Introduction   4th Edition
   By K. Binder and D. W. Heermann
81 **Core-Level Spectroscopy in Condensed Systems**
   Editors: J. Kanamori and A. Kotani
82 **Photoelectron Spectroscopy**
   Principle and Applications   2nd Edition
   By S. Hüfner
83 **Physics and Technology of Submicron Structures**
   Editors: H. Heinrich, G. Bauer, and F. Kuchar
84 **Beyond the Crystalline State**   An Emerging Perspective   By G. Venkataraman, D. Sahoo, and V. Balakrishnan
85 **The Quantum Hall Effects**
   Fractional and Integral   2nd Edition
   By T. Chakraborty and P. Pietiläinen
86 **The Quantum Statistics of Dynamic Processes**
   By E. Fick and G. Sauermann
87 **High Magnetic Fields in Semiconductor Physics II**
   Transport and Optics   Editor: G. Landwehr
88 **Organic Superconductors**   2nd Edition
   By T. Ishiguro, K. Yamaji, and G. Saito
89 **Strong Correlation and Superconductivity**
   Editors: H. Fukuyama, S. Maekawa, and A. P. Malozemoff

Springer Series in Solid-State Sciences
Editors: M. Cardona   P. Fulde   K. von Klitzing   H.-J. Queisser

Managing Editor: H. K. V. Lotsch

90 **Earlier and Recent Aspects of Superconductivity**
Editors: J. G. Bednorz and K. A. Müller

91 **Electronic Properties of Conjugated Polymers III** Basic Models and Applications
Editors: H. Kuzmany, M. Mehring, and S. Roth

92 **Physics and Engineering Applications of Magnetism** Editors: Y. Ishikawa and N. Miura

93 **Quasicrystals** Editors: T. Fujiwara and T. Ogawa

94 **Electronic Conduction in Oxides** 2nd Edition
By N. Tsuda, K. Nasu, F. Atsushi, and K.Siratori

95 **Electronic Materials**
A New Era in Materials Science
Editors: J. R. Chelikowsky and A. Franciosi

96 **Electron Liquids** 2nd Edition   By A. Isihara

97 **Localization and Confinement of Electrons in Semiconductors**
Editors: F. Kuchar, H. Heinrich, and G. Bauer

98 **Magnetism and the Electronic Structure of Crystals** By V.A. Gubanov, A.I. Liechtenstein, and A.V. Postnikov

99 **Electronic Properties of High-T_c Superconductors and Related Compounds**
Editors: H. Kuzmany, M. Mehring, and J. Fink

100 **Electron Correlations in Molecules and Solids**   3rd Edition   By P. Fulde

101 **High Magnetic Fields in Semiconductor Physics III** Quantum Hall Effect, Transport and Optics   By G. Landwehr

102 **Conjugated Conducting Polymers**
Editor: H. Kiess

103 **Molecular Dynamics Simulations**
Editor: F. Yonezawa

104 **Products of Random Matrices**
in Statistical Physics   By A. Crisanti, G. Paladin, and A. Vulpiani

105 **Self-Trapped Excitons**
2nd Edition   By K. S. Song and R. T. Williams

106 **Physics of High-Temperature Superconductors**
Editors: S. Maekawa and M. Sato

107 **Electronic Properties of Polymers**
Orientation and Dimensionality of Conjugated Systems   Editors: H. Kuzmany, M. Mehring, and S. Roth

108 **Site Symmetry in Crystals**
Theory and Applications   2nd Edition
By R. A. Evarestov and V. P. Smirnov

109 **Transport Phenomena in Mesoscopic Systems**   Editors: H. Fukuyama and T. Ando

110 **Superlattices and Other Heterostructures**
Symmetry and Optical Phenomena   2nd Edition
By E. L. Ivchenko and G. E. Pikus

111 **Low-Dimensional Electronic Systems**
New Concepts
Editors: G. Bauer, F. Kuchar, and H. Heinrich

112 **Phonon Scattering in Condensed Matter VII**
Editors: M. Meissner and R. O. Pohl

113 **Electronic Properties of High-T_c Superconductors**
Editors: H. Kuzmany, M. Mehring, and J. Fink

114 **Interatomic Potential and Structural Stability**
Editors: K. Terakura and H. Akai

115 **Ultrafast Spectroscopy of Semiconductors and Semiconductor Nanostructures**
2nd Edition   By J. Shah

116 **Electron Spectrum of Gapless Semiconductors**
By J. M. Tsidilkovski

117 **Electronic Properties of Fullerenes**
Editors: H. Kuzmany, J. Fink, M. Mehring, and S. Roth

118 **Correlation Effects in Low-Dimensional Electron Systems**
Editors: A. Okiji and N. Kawakami

119 **Spectroscopy of Mott Insulators and Correlated Metals**
Editors: A. Fujimori and Y. Tokura

120 **Optical Properties of III-V Semiconductors**
The Influence of Multi-Valley Band Structures
By H. Kalt

121 **Elementary Processes in Excitations and Reactions on Solid Surfaces**
Editors: A. Okiji, H. Kasai, and K. Makoshi

122 **Theory of Magnetism**
By K. Yosida

123 **Quantum Kinetics in Transport and Optics of Semiconductors**
By H. Haug and A.-P. Jauho

124 **Relaxations of Excited States and Photo-Induced Structural Phase Transitions**
Editor: K. Nasu

125 **Physics and Chemistry of Transition-Metal Oxides**
Editors: H. Fukuyama and N. Nagaosa

You are one click away from a world of physics information!

Come and visit Springer's
Physics Online Library

Books
- Search the Springer website catalogue
- Subscribe to our free alerting service for new books
- Look through the book series profiles

You want to order?   Email to: orders@springer.de

Journals
- Get abstracts, ToC´s free of charge to everyone
- Use our powerful search engine SpringerLink Search
- Subscribe to our free alerting service SpringerLink *Alert*
- Read full-text articles (available only to subscribers of the paper version of a journal)

You want to subscribe?   Email to: subscriptions@springer.de

Electronic Media
- Get more information on our software and CD-ROMs

You have a question on an electronic product?   Email to: helpdesk-em@springer.de

● Bookmark now:

http://www.springer.de/phys/

Springer · Customer Service
Haberstr. 7 · D-69126 Heidelberg, Germany
Tel: +49 (0) 6221 345 - 0 · Fax: +49 (0) 6221 345-4229
d&p · 006437_sf1c_1c

Druck: betz-druck GmbH, D-64291 Darmstadt
Verarbeitung: Buchbinderei Schäffer, D-67269 Grünstadt